Ceramic Nanomaterials and Nanotechnology II

Related titles published by The American Ceramic Society

Progress in Nanotechnology
©2002, ISBN 1-57498-168-4

Ceramic Nanomaterials and Nanotechnology (Ceramic Transactions Volume 137)
Edited by Michael Z. Hu and Mark R. De Guire
©2002, ISBN 1-57498-152-8

Other titles of interest

Innovative Processing and Synthesis of Ceramics, Glasses, and Composites VII (Ceramic Transactions Volume 154)
Edited by J.P. Singh and Narottam P. Bansal
©2003, ISBN 1-57498-208-7

Advances in Ceramic Matrix Composites VIII (Ceramic Transactions Volume 139)
Edited by J.P. Singh, Narottam P. Bansal, and M. Singh
©2002, ISBN 1-57498-154-4

Innovative Processing and Synthesis of Ceramics, Glasses, and Composites VI (Ceramic Transactions Volume 135)
Edited by Narottam P. Bansal and J.P. Singh
©2002, ISBN 1-57498-150-1

Innovative Processing and Synthesis of Ceramics, Glasses, and Composites V (Ceramic Transactions Volume 129)
Edited by Narottam P. Bansal and J.P. Singh
©2002, ISBN 1-57498-137-4

Innovative Processing and Synthesis of Ceramics, Glasses, and Composites IV (Ceramic Transactions Volume 115)
Edited by Narottam P. Bansal and J.P. Singh
©2000, ISBN 1-57498-111-0

Innovative Processing and Synthesis of Ceramics, Glasses, and Composites III (Ceramic Transactions Volume 108)
Edited by J.P. Singh, Narottam P. Bansal, and Koichi Niihara
©2000, ISBN 1-57498-095-5

Ceramic Nanomaterials and Nanotechnology II

Proceedings of the Nanostructured Materials and Nanotechnology symposium held at the 105th Annual Meeting of The American Ceramic Society, April 27–30, in Nashville, Tennessee

Edited by

Mark R. De Guire
Case Western Reserve University

Michael Z. Hu
Oak Ridge National Laboratory

Yury Gogotsi
Drexel University

Song Wei Lu
PPG Industries, Inc.

Published by
The American Ceramic Society
735 Ceramic Place
Westerville, Ohio 43081
www.ceramics.org

Proceedings of the Nanostructured Materials and Nanotechnology symposium held at the 105th Annual Meeting of The American Ceramic Society, April 27–30, 2003, in Nashville, Tennessee

COVER PHOTO: SEM fracture image of a carbon nanotubes-Fe-Al_2O_3 hot-pressed composite" is courtesy of E. Flahaut, S. Rul, F. Lefévre-Schlick, Ch. Laurent, and A. Peigney and appears as figure 3 in their paper "Carbon Nanotubes-Ceramic Composite" which begins on page 71.

For information on ordering titles published by The American Ceramic Society, or to request a publications catalog, please call 614-794-5890.

4 3 2 1–07 06 05 04

ISSN 1042-1122
ISBN 1-57498-203-6

Contents

Preface

This volume of Ceramic Transactions consists of papers presented at Symposium 7 on Nanostructured Materials and Nanotechnology, which was held during the 105th Annual Meeting of The American Ceramic Society in Nashville, Tennessee, April 27-30, 2003. A total of 74 papers were presented in the symposium: 55 oral presentations (including 17 invited papers) and 19 posters, spanning the full three days of the meeting. Reflecting the truly international character of the symposium and of nanomaterials research, this compilation contains papers from authors working in eleven countries.

This symposium followed two successful symposia on nanostructured materials at the 2001 and 2002 Annual Meetings. The 2003 symposium consisted of sessions on: the synthesis and processing of ceramic nanomaterials; nanoparticles and particle assemblies, including photonic structures; the fabrication and properties of ceramic-ceramic, ceramic-metal, and ceramic-polymer nanocomposites; nanotubes of carbon, boron nitride, and other ceramics; characterization and properties of nanomaterials; and industrial development and applications of nanomaterials. Throughout the symposium, both the quality of the presentations and the interest of the attendees were high.

In a relatively short time, the field of nanostructured materials has expanded from a novel area of research to a technology with a significant and rapidly growing commercial sector. In recognition of this development, the symposium included an entire session (three hours of talks, plus a one-hour panel discussion) on commercial aspects of nanotechnology.

Nineteen of the papers presented at the symposium, plus a summary of the panel discussion, are contained in this volume. Together they offer wide-ranging coverage of the current state of nanomaterials science and technology. The papers are organized into four chapters: Synthesis and Processing of Nanoparticles and Nanostructured Assemblies; Fabrication and Properties of Nanocomposites; Characterization and Properties of Nanomaterials; and Industrial Development and Applications of Nanomaterials.

The symposium was sponsored by The American Ceramic Society and the Basic Science Division of ACerS, and by the NSF Particulate Materials

Center (Prof. James Adair, director) at the Pennsylvania State University. The editors would like to thank Greg Geiger at ACerS Headquarters for invaluable assistance in organizing the review process and coordinating the production of this proceedings volume. Finally, the editors thank all of the authors who contributed manuscripts to the proceedings or who assisted with reviewing manuscripts for this volume.

Mark R. De Guire

Michael Hu

Yury Gogotsi

Song Wei Lu

Synthesis and Processing of Nanoparticles and Nanostructured Assemblies

Structural and Optical Properties of CdSe Nanoparticles Prepared by Mechanical and Chemical Alloying

G. L. Tan [a)], U. Hommerich [b)], D. Temple [b)], G. Loutts [c)]

a) Dept. of Materials Science & Engineering, University of Pennsylvania, 3231 Walnut Street, Philadelphia, PA19104.

b) Physics Department, Hampton University, Hampton, VA23668

c) Center for photonic Materials Research, Norfolk State University, 700 Park Avenue, Norfolk, VA 23504

Abstract

CdSe nanocrystals have been successfully synthesized by both physical method, ball milling of elemental Cd and Se powders and chemical method. Compared to chemical methods, the advantage of this physical approach is being able to achieve quantum confinement effect of uncapped CdSe nanocrystals, and furthermore to investigate the influence of capping surface effects on the optical properties by replacing different capping organic ligands. XRD pattern revealed that initial products of CdSe nanocrystals with wurtzite structure had been synthesized through mechanical alloying a mixture of elemental Cd and Se powders for several hours. A phase transition from wurtzite structure to cubic zinc blende structure was observed upon prolonging ball milling time above 5 hours. The as-milled powders exhibit dark red colorization after several hours milling but their dispersion in non-polarizing organic ligands did not show any colorization. After capping with polarizing organic ligands, the capped CdSe nanocrystals dispersed in hexane solution showed red color with the absorption peaks locating within the wavelength range from 400-600 nm, which corresponds to the band gap energy of capped CdSe nanocrystals in the polarizing organic ligands.

1. Intro*duction*

Basic spectroscopic and structural studies on differently prepared CdSe nanoparticles have received increasing interest during the past ten years[1,2]. Quantum dots (QD) provide an opportunity to investigate the behavior of semiconductors in the finite size regime[3]. Therefore, the chemical synthesis and the properties of highly luminescent II-VI semiconductor nanoparticles have been extensively investigated allowing from pure basic research to the application of these nanomaterials in electrical and opt-electric devices[4,5,6,7,8]. Most of the studies involved cadmium serenade nanocrystals, which can be prepared either in aqueous solution using thiol as the stabilizing agents[9,10,11,12,13,14], or in high boiling coordinating solvents[15,16]. In the early 90s, high quality CdSe nanocrystals became available by using $Cd(CH_3)_2$ as the cadmium precursor and the technical grade trioctylphosphine oxide (Tech TOPO) as the reaction solvent[15,17]. The synthesis of CdSe and CdTe nanocrystals through this organometallic route has led the synthesis of high quality semiconductor nanocrystals for about 10 years[18]. However some key chemicals used in this traditional route are extremely toxic, pyrophoric, explosive and expensive. After that, during early 20s, different kinds of safe, low cost inorganic compounds of CdO, $Cd(AC)_2$, and $Cd(CO_3)$ had been used as cadmium precursor, instead of toxic, explosive and expensive $Cd(CH_3)_2$, for the synthesis of high quality CdTe and CdSe nanocrystals[19,20]. The high boiling solvent was extended to Stearic acids; fatty acids, amines, phosphine oxides and phosphine acids. Fundamental spectroscopic and structural studies on CdSe nanoparticles by various synthesis routes received an increasing interest during the past several years[21,22].

Since the pioneering work of Efros[23] and Brus[24] devoted to the size -quantization in semiconductor nanoparticles, the research on nanostructures has become a flourishing field in chemistry, physics and material science[25,26]. An important landmark in the development of wet chemical routes for Cadmium sulphide nanoparticles was, together with the non-aqueous TOP/TOPO technich[6], the use of thiols as stabilizing agents in aqueous solutions. All above reports mainly concentrated on the investigation for the quantum dots fabricated by chemical process, in which capped polarization organic ligands are coated on the surface of the quantum surface to keep the dots dispersing in the host materials. Here we report a novel approach to fabricate uncapped CdSe quantum dots by a physical method, ball milling process. The potential advantage of the physical approach is the possibility of achieving the intrinsic quantum confinement effect of the uncapped CdSe nanocrystals excluding the influence of the organic surface effect, i.e. the interaction between the surfaces of the nanocrystals and capped organic ligand, as well as the possible batch production of the nanocrystals in larger scale compared with chemical method. The influence of the surface on QD properties is complicated and remains controversial. So reducing this influence can be valuable in determining intrinsic QD's physical properties.

2. EXPERIMENT

The starting materials were high purity cadmium (99.99%) and Selenium (99.999%) elemental powders. Mechanical alloyed CdSe particles were prepared by ball milling stoichiometric amounts of Cd and Se elemental powders with atomic ratio of 1:1. The starting powders were sealed in a hardened steel vial with hardened steel balls. Milling balls with different diameters (2-12 mm) were used.

Milling was performed on a SPEX 8000D Mix/miller using a ball to powder mass ratio 20:1. Small amounts of as-milled powders were taken out of the vial within different time intervals for structural and optical measurements. The structural phase composition of as milled powders for different period was detected by Rigaku powder X-ray diffractometer. The mean crystallite diameter was determined from the peak width of x-ray diffraction spectra using the Scherer formula. The optical measurement was carried out using uncapped CdTe nanocrystals and capped CdTe nanocrystals dispersed in hexane.

3. RESULTS AND DISCUSSION

3.1. Structural Characterization

Cadmium and selenium elemental powders were mixed together in stainless steel vials and mounted on a SPEX 8000D ball-milling mix/miller. CdSe nanoparticles were synthesized via the reaction of $Cd + Se \rightarrow CdSe$ under high frequency impact of hardened alloy balls on the wall. The crystal structure of the mechanically alloyed products was identified using an X-ray diffractometer.

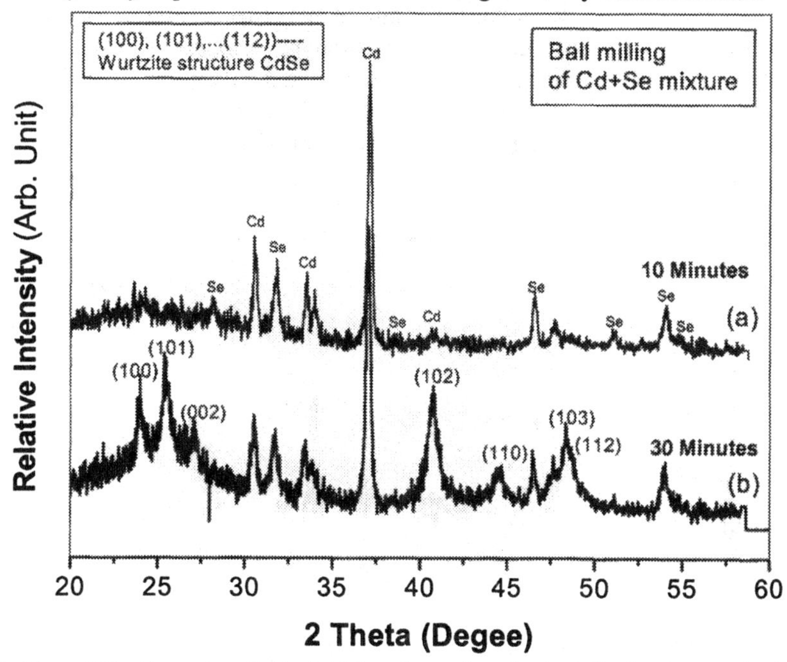

Figure 1 , X-ray diffraction pattern for the initial products of mechanical alloying elemental Cd & Se mixture powders within 30 minutes.

Figure 1 shows X-ray diffraction pattern for the mixture of elemental Cd and Se powders and their products of mechanochemical reaction during the early ball milling process. In the initial stage of the ball milling process, i.e., 10 minutes, no evident diffraction peaks of any type compounds were observed except those indexed from the mixture of elemental Cd and Se powder as shown in Figure 1 (a).

While small amount of CdSe with wurtzite structure were successfully fabricated by mechanical alloying process for only 30 minutes, mixing with the majority phases of Cd and Se elemental powders being indexed in the XRD pattern. It can be seen that after ball milling for 30 minutes, wurtzite CdSe diffraction peaks ((100), (101), (002), (102), (110), (103), (112)) clearly showed up in the XRD pattern (Figure 1 (b)), coexisting with the other phases of elemental powders of Cd and Se. The intensity ratio of the compound phase to elemental phases was estimated to be 0.75. The mechanical alloying reaction was completed as $Cd + Se \rightarrow CdSe$. Under normal synthesis condition (i.e. without milling), the reaction mentioned above will not happen at room temperature. During ball milling process, this self propagating reaction could take place if the contribution of the high frequent milling impact energy from the ball mill machine to the energy of formation (ΔG) for this reaction is big enough that the entropy of the reaction becomes positive ($\Delta S > 0$)[27]. Due to the extra high frequent impacting of the ball on mixture powders against the wall of the vials, so the impacting energy from the milling machine should be big enough to initiate the reaction above by changing the entropy to be positive, like mechanical alloying process of other inter-metallic compounds[27]. Once the reaction milling proceeds, the grain size of the powders becomes smaller and the specific surface area becomes larger, namely the surface density decreases and the impacting energy transferred to the powders increases, causing the enhancement of the entropy and formation of CdSe compounds through mechanical alloying process.

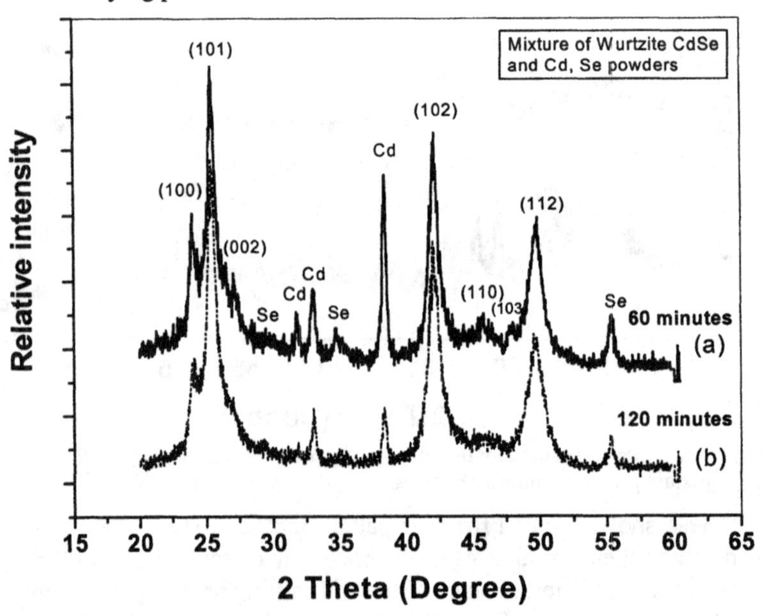

Figure 2 X-ray diffraction pattern for early products of mechanical alloying elemental Cd & Se mixture powders between 60 minutes and 120 minutes, showing the formation of wurtzite structural CdSe compound.

With extension of ball milling time to 60 minutes (Figure 2 (a)), Diffraction signals from elemental powders become much weaker while the diffraction intensity of Wurtzite structural CdSe compound began to dominate mixture powder, which is still composed of elemental Cd and Se powders as well as alloyed CdSe compound powders. The diffraction intensity ratio of the compound phase to elemental phase increases to be about 1.38. Further prolonging the ball milling time to 120 minutes did not lead to formation of additional compounds but increases the diffraction intensity ratio of CdSe compound to elemental Cd phase up to 4.16 as shown in Figure 2 (b), indicating that most of elemental powders have transferred to CdSe compound through mechanical allying during ball milling process, with few Cd and Se powders being non-reacted in the mixture. Wurtzite structural CdSe particles are the main products during the early stage of ball milling process (from begin to 2 hours).

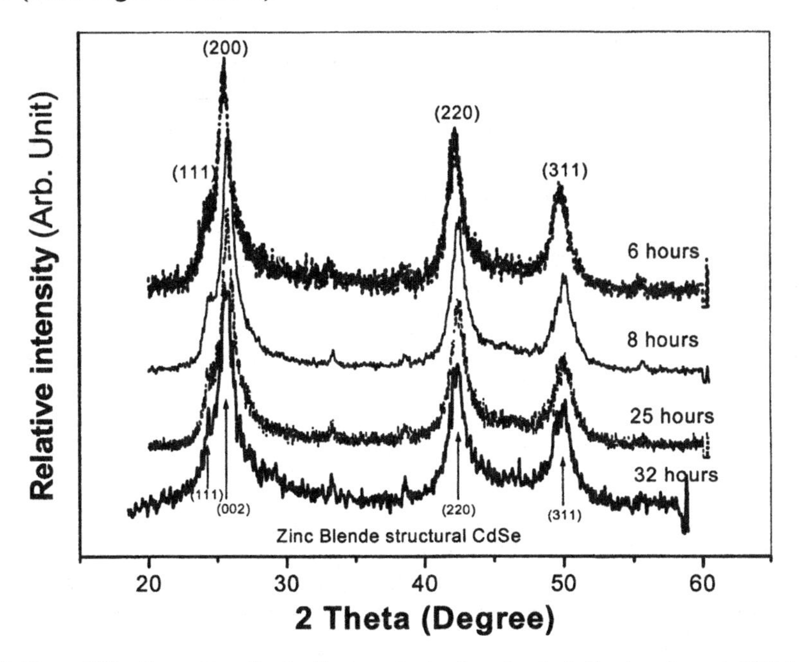

Figure 3 X-ray diffraction pattern for the final products of mechanical alloying elemental Cd & Se mixture powders within 6 hours and 32 hours, showing the phase transition of CdSe compound from Hexagonal wurtzite structure to cubic Zinc blende structure during ball milling process upon long milling period for above 5 hours.

When the mechanical alloying process continued further longer, the concentration of elemental powders in the mixture becomes less and less, in other word the diffraction intensity of elemental powders becomes smaller and smaller until finally they disappeared after ball milling for up to 32 hs, as shown in Figure 3. It can be also seen from Figure 3 that within the period of ball milling for above 5 hours, a phase transition for CdSe powders from wurtzite structure to cubic zinc blende structure had occurred. The wurtzite structure of the CdSe compound being synthesized during the early mechanical alloying stage, i. e. within first two hours, transferred into cubic zinc blende structure after further ball milling for above 6 hours. This type of cubic CdSe compound remains to be the stable phase even if ball milling process was further carried out for up to 32 hours. Four main diffraction peaks corresponding to the {111}, {002}, {220} and {311} crystal planes of cubic CdSe structure had been observed in all products of the process of ball milling for 6 hours, 8 hours, 25 hours and 32 hours as shown in Figure 3. The phase transformation from wurtzite structure to rock salt structure for CdSe nanocrystals at approximately 6 Gpa was observed by S. H. Tolbert and A. P. Alivisatos[28], but not observed in our case due to much lower pressure impacting on the powders. Instead we observed the transformation to zinc blende structure after several hours milling process. Single phase CdSe nanocrystals with cubic zinc blende structure were formed after ball milling for more than 8 hours. So cubic structural CdSe in single phase keeps itself to be the final product of the mechanical alloying process no matter how long the ball milling is. The amount of other structural phases and elemental powders are so small that they are not enough to be detected in the XRD pattern. Within the ball milling period of 6 hours to 32 hours, no obvious additional change like the phase transition was observed, only resulting in the XRD CdSe peaks broadening and the full width at half maximum (FWHM) of the peaks increases with ball milling time. The broadening effect of the peaks arises from small size particles, indicating that the particle size decreases with longer ball milling time. Furthermore, the {111} peak was rescanned at a slow angle scanning process (0.002 degree/step) to improve peak quality, in order to estimate the average size of the particles. It is found that the FWHM from the sample ball-milled for 32 hours is much wider than that for 2 hours, indicating that the former is consisted of much smaller particles than the latter. According the Scherer equation, the average particle sizes of the two samples were estimated to be 8 nm and 27 nm, respectively.

3.3. Optical Properties of CdSe nanoparticles through mechanical alloying

The as-milled powders were taken out of the vials within different intervals in order to evaluate their optical properties. The as-milled powders exhibit dark red colorization after several hour ball milling, which is quite different from the black powders of CdSe nanoparticles prepared by the same milling process as described somewhere else. The powders were first dispersed in the non-polarizing organic solution hexane, which did not show any colorization. The colorless CdSe colloid dispersion reveals that their absorption peak could locate within ultraviolet wavelength range instead of visible wavelength range, which is possibly due to the fact that only extremely small particles can float in the solution because floating force merely arises from the solution and is not big enough to support big CdSe particles. In contrast, after capping with polarizing organic ligands like TOP/TOPO, the capped CdSe nanocrystals dispersed in hexane solution showed red color with absorption peaks locating within the wavelength range from 400-600 nm as shown in Figure 4. In later case, the charge interaction force between the capping polarization organic ligand on the particle surface can support the bigger capped CdSe nanoparticles to float in the solution. These big CdSe nanoparticles have smaller band gap energy, which absorbs and emits light within visible wavelength range instead of ultraviolet wavelength range, making the dispersion solution exhibit red color.

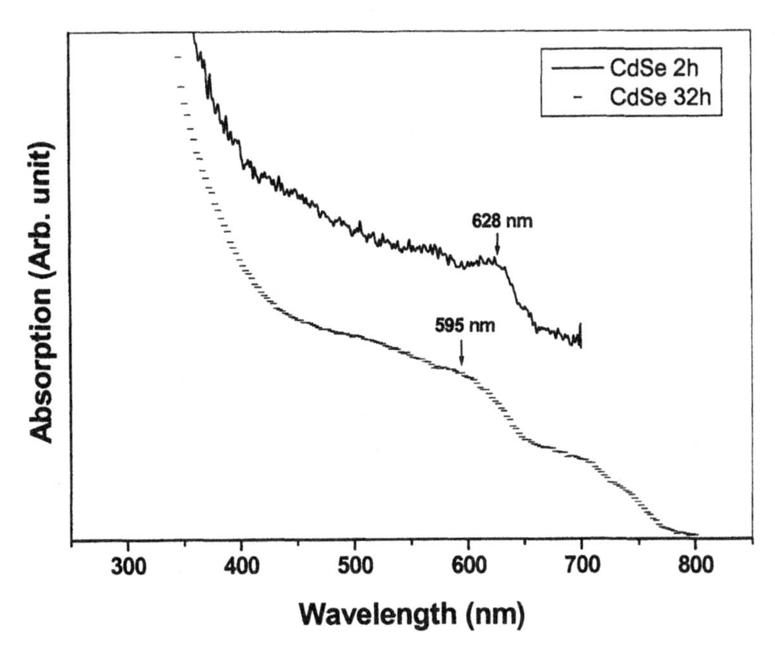

Figure 4 UV visible absorption spectrum for the colloid CdSe nanoparticles in hexane, the nanoparticles was fabricated by mechanical alloying for 2 hs and 32 hs respectively.

The visible absorption spectrum of the dispersion solution was shown in Figure 4. It can be seen that the colloid dispersion solution containing CdSe nanoparticles as milled for 2 hs exhibit a small absorption peak at 626 nm (Figure 4 (a)). As we can see from Figure 2 that the as-milled CdSe powders for 2 hours have wurtzite structure, whose bulk band gap is 1.751 eV (708nm). The energy shift for the band gap of CdSe nanocrystals compared to bulk value can be expressed by Brus's formula, which was induced by quantum size confinement effect when the particle size is close to its Bohr excite radius:

$$\Delta E = \frac{\hbar^2 \pi^2}{2}\left(\frac{1}{m_e} + \frac{1}{m_h}\right)\frac{1}{R^2} - \frac{1.8e^2}{\varepsilon_2}\cdot\frac{1}{R} + \frac{e^2}{R}\sum_{n=1}^{\infty}\alpha_n\left(\frac{S}{R}\right)^{2n} \quad \text{......(1)}$$

A blue shift of 0.23 eV for the band gap energy of 2h as-milled CdSe nanoparticles compared to its wurtzite bulk value (from 708 nm to 626 nm) was observed in Figure 4, revealing that 2 hours ball milling process already produced very small CdSe particles in nanometer scale being synthesized through mechanical alloying process. This blue shift of band gap energy can be calculated from equation 1, resulting in corresponding particle size of 6.5 nm for CdSe nanocrystals.

The disperse solution containing cubic CdSe nanocrystals as milled for 32 hours exhibits a wide absorption platform (from 400nm to 700nm) with two absorption edges at 595 nm and 704 nm respectively. This result reveals a wide size distribution of cubic CdSe nanoparticles in the solution. As we can see from Figure 3 that the 32h as-milled CdSe powders have zinc blende structure, whose bulk band gap is not available. Because it has the same structure and very close lattice parameters as zinc blende CdTe, they may have similar band gap value of 1.475 eV (841 nm). Under this assumption, a much bigger blue shift of 2.9 eV for the band gap energy of 32h as-milled CdSe nanoparticles compared to its cubic bulk value (from 841 nm to 400nm) could have been observed in Figure 4 (b), elucidating that 32 hours ball milling process already produced extremely small CdSe particles in several nanometers scale being fabricated through ball milling process. The big blue shift of 2.9 eV for band gap energy of cubic CdSe nanoparticles can also be calculated from equation 1, indicating that the shortest absorption edge at around 400 nm is corresponding to CdSe nanocrystals with 1.6 nm particle size. The absorption edges at 595nm and 704nm in Figure 4 (b) correspond to the colloid dispersion of CdSe nanocrystals with particle size of 6 nm and 11 nm respectively. Bigger particles could not disperse in the solution by floating force even if they are capped by TOP/TOPO on the surface. These large particles therefore deposit onto the bottom and ply no role on their optical properties. A blue shift of the absorption peak or absorption edge could also be observed between 32h sample and 2 h samples, implying that longer ball milling time produced smaller CdSe nanoparticles with bigger band gap energy.

3.3. Optical Properties of CdSe nanoparticles by chemical alloying

In order to compare optical properties of CdSe nanoparticles prepared by mechanical alloying process with that prepared by chemical alloying method, we had also synthesized CdSe nanoparticles by chemical approach using the method widely reported[15,18,20]. After dissolving these TOPO capped CdSe nanoparticles in acetone, a small amount of PMMA solution was added into the colloid dispersion solution. The mixture of the CdSe nanoparticles dispersion and organic PMMA solution exhibits fresh red color. The UV visible absorption spectrum of the mixture solution is shown in Figure 5. The absorption peaks for both spectrum curves corresponding to two kinds of solvents of (a) Methanol and (b) Hexane respectively, in which CdSe nanoparticles were dissolved, behave quite differently in shape positions. The UV visible spectrum for CdSe-methanol shows up two absorption peaks at 270 nm and 409 nm respectively and that for CdSe-Hexane shows up only one peak at around 375 nm. The band gap energy of the CdSe nanoparticles corresponding to these peaks (4.6 eV, 3.03eV and 3.31 eV respectively) is much bigger than its bulk value at 1.751 eV, indicating that the particle size is extremely small, i.e.1.2-3 nm. This big blue shift of the band gap energy for CdSe nanoparticles prepared by chemical method is seldom seen in other literatures[5,12,15], but similarly phenomenon had been observed in the UV visible spectrum of uncapped CdTe nanoparticles prepared by mechanical alloying process[29]. The difference between the two systems comes from the surface charge contribution due to the different dielectric function of two solvents, as being revealed by the S factor in the last term of equation 1.

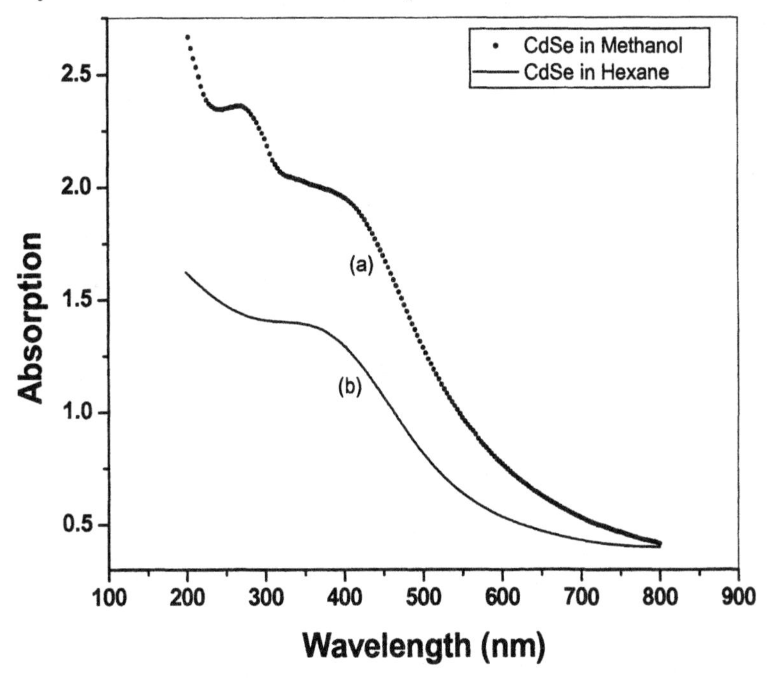

Figure 5 UV visible absorption spectrum of the colloid dispersion of CdSe nanoparticles in hexane, the nanoparticles had been synthesized by chemical alloying process

CONCLUSION

High quality CdSe nanoparticles were synthesized mainly by mechanical alloying and also by chemical alloying process. During ball milling process, wurtzite structural CdSe powders were the initial product within the first two hours. A phase transition from wurtzite structure to cubic zinc blende occurred upon milling for above 4 hours. The cubic phase remains stable to keep itself to be the final product for long period ball milling process. Optical properties exhibit wide absorption edge within 420nm to 750nm, revealing a wide size distribution range. A big blue shift (2.85eV) of the band gap energy for CdSe nanoparticles prepared by chemical alloying process was observed. Surface charge effect of different solvents on nanoparticles optical properties was observed in chemical alloyed samples, which was caused by different dielectric function of solvents.

ACKNOWLEDGMENT
The work at Hampton University was supported by NASA through grant NCC-1-125.

References

1 C. B. Murray, D. J. Norris, M. G. Bawendi, J. Am. Ceram. Soc. 1993, 115, 8706.
2 U. Resch, H. Weller, A. Henglein, Langmuir 1989, 5, 1015.
3 M. Nirmal and L. Brus, Acc. Chem. Res. 1999, 32, 407.
4 Brus, L. E., Appl. Phys. A. 1991, 53, 465. Weller, H., Adv. Mater. 1993, 5, 88. Alivisatos, A. P. , J. Phys. Chem. 1996, 100, 13226. Woggon, U., Optical Properties of Semiconductor Quantum Dots; Springer-Verlag: Berlin, 1997. Gapnonko, S. V. , Optical Propertiesof Semiconductor Nanocrystals, Cambridge University Press: Cambridge, 1998.
5 Colvin, V. L.; Schlamp, M. C.; Alivisatos, A. P.; McEuen, P. L.; Nature 1997, 370, 354.
6 Klien, D. L.; Roth, R.; Lim, A. K. L.; Alivisatos, A. P.; McEuen, P. L.; Nature 1997, 389, 699
7 Ridley, B. A.; Nivi, B.; Jacobson, J. M.; Science 1999, 286, 746.
8 Kim, S. H.; Medeiros-Rebeiro, G.; Ohlbergf, D. A. A.; Stanley Williams, R.; Heath, J. R.; j. Phys. Chem. B. 1999, 103, 10341.
9 Herron, N.; Calabrese, J. C.; Farneth, E. E.; Wang, Y.; Science 1993, 259, 1426.
10 Rajh, T.; Mici, O. I.; Nozik, A.. J.; J. Phys. Chem. 1993, 97, 11999.
11 Vossmeyer, T.; Katsikas, L.; Giersig, M.; Popovic, I. G.; Diesner, K.; Chemseddine, A.; Eychmueller, A.; Weller, H.; J. Phys. Chem., 1994, 98, 7665.
12 Rogach, A. L.; Katsikas, L.; Kornowski, A.; Su, D.; Eychmueller, A.; Weller, H.; Ber. Bunsen-Ges. Phys. Chem. 1996, 100, 1772, 1997, 101, 168.
13 Rogach,A; Kornowski, A.; Gao, M.; Eychmueller, A.; Weller, H.; J. phys. Chem. B 1999, 10, 3065.
14 N. N. Mamedova, N. A. Kotov, A. L. Rogach, J. Studer; Nano. Lett., 1, 281, 2001.
15 Murrray, C. B.; Norris, D. J.; Bawendi, M. G.; J. Am. Chem. Soc. 1993, 115, 8707.
16 Boween Katari, J. E.; Colvin, V. L.; Alivisatos, A. P.; J. Phys. Chem. 1994, 98, 4109.
17 Brennan, J. G.; Siegrist, T.; Gaeeol, P. J.; Stucznski, S. M.; Reynders, P.; Brus. L. E.; Steigerwald, M. L. , Chem. Matter 1990, 2, 403.
18 Peng, X. G.; Mannam, L.; Yang, W. D.; Wickham, J.; Scher, E.; Kadacanich, A.; Alisvisatos, A. P.; Nature 2000, 404, 59-61.
19 Qu, L.; Peng, Z. A.; Peng, X.; Nano. Lett. 0, A, 2001.
20 Peng, X.; Manna, Yang, W.; Wickham, J.; Scher, E.; Kadacanich, A.; Alisvisatos, A. P.; Nature 2000, 404, 59-61.
21 Resch U.; Weller H.; Henglei H.; Langmuir 1989, 5, 1015.
22 Rajh T.; Micic, O. I.; Nozik, J.; J. Phys. Chem. 1993, 97, 1999.
23 Efros, Al. L.; Efros, A. L.; Sov. Phys. Semicond. 16, 772, 1982.
24 Brus, L. E.; J. Chem. Phys. 80, 4403, 1984.
25 Weller, H.; Angew. Chem. Int. Ed. Engl. 32, 41, 1993.
26 Alivisatos, A. P.; J. Phys. Chem. 100, 13226, 1996.
27 Tan, G. L.; Wu, X. J.; Zhao, M. H.; Zhang, H. F; J. Mater. Sci. 35, 3151, 2000.
28 Tolbert, A. H.; Alivisatos, A. P.; J. Chem. Phys. 102 (11), 4642, 1995.
29 Tan G. L. ; Hommerich, U.; Temple, D.; Wu, N. Q.; Zhang, J. G.; Louts, G.; Scripta Materialia, 48, 1469, 2003.

NANO-SIZED SILICON NITRIDE POWDER SYNTHESIS VIA AMMONOLYSIS OF SiO VAPOR

Panut Vongpayabal and Shoichi Kimura*
Department of Chemical Engineering
Oregon State University
Corvallis, OR 97331-2702

ABSTRACT

An 89 mm-diameter vertical tubular-flow reactor was built to study the kinetics of nano-sized silicon nitride powder synthesis via the ammonolysis of SiO vapor at temperatures ranging from 1300°C to 1400°C. When the supply of NH_3 was in great excess over SiO, the rate of nano-sized powder synthesis was independent of NH_3 concentration and of first order with respect to SiO concentration. The apparent activation energy of pseudo-first order rate constant was 180 kJ/mol. The locations of whisker formation in the reactor were identical to those of flow stagnation predicted by computational fluid dynamics. Measurements of whisker formation rates were correlated in terms of a power law expression to simulation-predicted reactant-gas concentrations.

INTRODUCTION

Over the past decades, silicon nitride (Si_3N_4) has been known as an excellent material for high-temperature structural applications due to its outstanding physical and mechanical properties under severe environment.[1] Beyond such properties of parts made from regular sub-micron silicon nitride powder, it is believed that nano-sized silicon nitride powder enhances additive-free sinterabilities and increases plasticity, permitting large plastic deformation at low temperature, as well as strength and toughness of sintered parts.[2]

The ammonolysis of silicon monoxide has recently been proposed, which uses common chemicals to produce nano-sized silicon nitride powder at low cost.[3]

$$3SiO(g) + 4NH_3(g) \longrightarrow Si_3N_4(s) + 3H_2O(g) + 3H_2(g)$$

This process has been proved, using a horizontal tubular-flow reactor, with a maximum nano-sized silicon nitride powder yield of 43% due to the formation of silicon nitride whiskers and crystals. This research further investigates factors that affect the synthesis of nano-sized silicon nitride powder via the ammonolysis of SiO vapor, including reactant concentrations, reaction temperature, and mixing-modes of reactants. A vertical tubular-flow reactor was built, which allowed flexible arrangements for supplying the two reactants, and the reaction kinetics as well as product morphologies were investigated using different methods for feeding gases. Reaction rates of nano-sized powder synthesis were measured at temperatures ranging from 1300°C to 1400°C.

To better understand the performance of reactor, a mathematical model was solved using a computational fluid dynamic program. The results of simulation were used for identifying correlations between whisker formation and gas flow in the reactor. An attempt was made to develop a kinetic expression for whisker formation based on estimated concentration profiles of reactant gases.

EXPERIMENT

Reactor Set-Up

Figure 1 illustrates a vertical tubular flow reactor used in this study. NH_3 was supplied through 1.4 mm ID alumina tube. SiO vapor was generated directly from SiO particles (purity 99.8%, 400 μm averaged size, Alfa Aesar) placed inside an SiO generator, a 7.13 mm ID × 90 mm L alumina tube. Both were placed in the 200 mm long uniform temperature zone, in which the temperature variation was within ±5°C. In each run, a prescribed amount of SiO particles, 0.6-3.7 g, were placed in the SiO generator kept in the uniform-temperature zone.

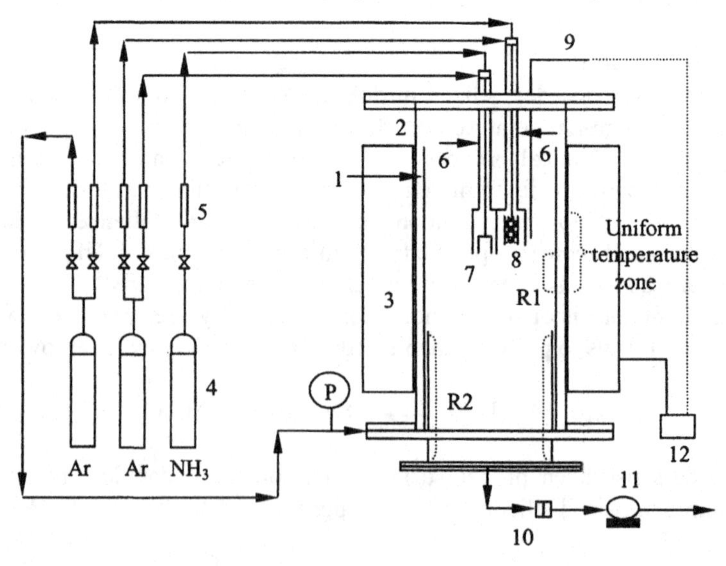

Figure 1. Experimental set-up: (1) reactor tube, (2) furnace tube, (3) furnace, (4) cylinders, (5) flow meters, (6) annular tubes, (7) NH_3 feeder, (8) SiO generator, (9) thermocouple, (10) filters, (11) vacuum pump, (12) temperature controller

Two sets of dual-concentric tubes, one outside the SiO generator and the other outside the NH_3 feeder, were used for supplying argon gas, surrounding respective SiO and NH_3 feeds, so that argon gas-curtain could act as diffusion barriers to prevent SiO vapor and NH_3 gas from reacting at the feeder outlets. The outer mullite tubes of NH_3 feeder and SiO generator were placed in contact side by side. Additional argon was also supplied as sheath gas along the reactor tube to prevent the back-flow of reactant gases and minimize powder attachment to the reactor wall. The total flow rate of gases was maintained about 52.5 lit/min at pressure slightly higher than atmospheric pressure. The reaction time was set to be 30 min, during which the SiO generation rate was roughly constant.[3]

Before starting the SiO-NH_3 reaction, the system was heated to a prescribed temperature with continuous flow of argon in the reactor tube outside the SiO generator and NH_3 feeder. When the system had reached the prescribed temperature, the reaction was initiated by supplying argon gas through the SiO generator. Simultaneously, NH_3 gas was introduced into the reacting zone through the NH_3 feeder. In each run, the molar feed ratio of NH_3/SiO at the feeder outlets was maintained in large excess of NH_3 over the stoichiometric ratio, ranging from about 100 to 1200 mol NH_3/mol SiO.

To collect silicon nitride powder generated in the vapor phase, filter papers were placed in the exhaust line downstream the reactor exit. After operation for 30 min, the reaction was terminated by stopping the argon gas supply through the SiO generator. The reactor was then cooled down to room temperature under the continuous flow of argon gas through the reactor tube. All silicon nitride products obtained at different locations inside the reactor were collected individually and weighed before further analysis. Product powder was observed using a Philips CM-12 scanning transmission electron microscope (TEM). The particle size distributions were measured by photon correlation spectroscopy (PCS), with a Beckman Coulter N4 Plus, and average particle sizes were determined.

The content of SiO remaining unreacted in product powder was determined by thermal treatment that allowed SiO to sublimate from the powder. Samples of product powder (about 10 mg) were heated in an argon stream at 1350°C for 1 hr, at which SiO possibly remaining in the powder samples sublimates. It has been known that silicon nitride powder formed through the ammonolysis of SiO vapor is amorphous and easily oxidized to silicon oxy-nitride (Si_2N_2O) when exposed to the atmosphere.[3-6] It was thus assumed that all product silicon nitride powder had been oxidized to silicon oxy-nitride before heating and remained unchanged upon heating at 1350°C for 1 hr. Fractions of SiO remaining unreacted were determined based on decreases in mass of samples due to the sublimation of SiO. The details for the experimental apparatus should be referred to elsewhere.[3]

RESULTS AND DISCUSSION

Product Morphologies

Two different morphologies of silicon nitride products were obtained at different locations in the reactor: (1) whiskers at the SiO generator outlet as well as on the reactor wall, region R1; and (2) fine powder on the reactor wall, region R2, as well as on the reactor bottom flange in addition to powder collected with external in-line filters. As illustrated in Figure 1, region R1 is 5-140 mm long inside the 200 mm-long uniform temperature zone, while region R2 is outside the heating zone, where the temperature abruptly drops below 1000°C. Figure 2 shows a TEM picture of Si_3N_4 powder collected with a filter, composed of particles with diameters ranging from 5 nm to 70 nm. The average diameter of particles shown in Figure 2 is 18 nm, measured by PCS. Powder collected in region R2 indicated roughly the same results.

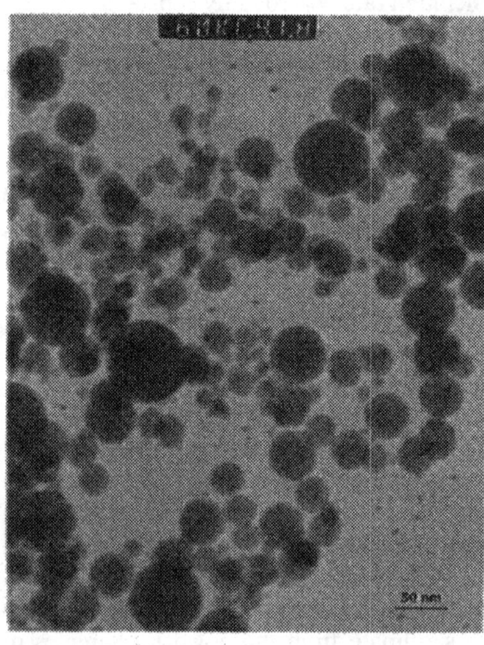

Figure 2. TEM picture of powder product.

Production Rate of Nano-Sized Powder

Dependency on NH_3 Feed Rate: The supply of NH_3 was in large excess of stoichiometric ratio over SiO. Table I compares the amount of nano-sized silicon nitride powder as well as the extent of SiO conversion for different NH_3/SiO feed ratios, while the feed rate of SiO is maintained roughly constant. Both the amount of nano-sized powder and the SiO conversion are independent of NH_3/SiO feed ratios covering a one-order-of-magnitude range. It is indicated that the SiO-NH_3 reaction is independent of the NH_3 concentration when the molar NH_3/SiO feed ratio is maintained in the range investigated.

Table I. Effects of NH_3 feed rates on the SiO-NH_3 reaction.

SiO feed rate $\times 10^3$ mmol/s	NH_3/SiO feed ratio mmol/mmol	Nano-sized powder, mmol	SiO conversion, X_{SiO}, -
1.72	98	0.53	0.895
1.65	200	0.52	0.849
1.57	488	0.56	0.891
1.57	976	0.54	0.828

Dependency on SiO Concentration: The performance equation for a plug flow reactor, in which the residence time of reactant is maintained constant and the temperature distribution is fixed, indicates that the extent of reactant conversion becomes independent of reactant feed concentration when the reaction is of first order with respect to the reactant concentration regardless the temperature distribution in the reactor.[7] The only exception is an instantaneous reaction in which the extent of conversion is always 100%.

Table II shows the extents of conversion of SiO obtained at different SiO feed rates. Because the total flow rate of gas is maintained roughly constant, the changes in the SiO feed rate correspond to the changes in the SiO concentration. The results shown in Table II indicate that the extent of SiO conversion is roughly constant, independently of SiO concentration. Obviously the extent of conversion is not 1.0, and hence the reaction is not instantaneous. Because the reaction is independent of NH_3 concentration when it is in large excess, the reaction is of first order with respect to the SiO concentration.

Table II. Effects of SiO feed rates on the SiO-NH_3 reaction.

SiO feed rate $\times 10^3$ mmol/s	NH_3/SiO feed ratio mmol/mmol	Fraction of powder, wt%	SiO conversion, X_{SiO}, -
0.63	1212	94.1	0.930
1.20	641	94.0	0.917
1.57	488	98.0	0.891
1.60	480	98.1	0.880
1.76	311	88.1	0.924

Rate constant: As stated earlier, a 200 mm long zone in the furnace is identified as the uniform temperature zone, in which temperature variation is within ±5°C, and outside of this zone may be defined as the variable temperature zone, in which temperature sharply drops. However, it is natural to consider that the reaction between SiO and NH_3 also takes place to some extent in the variable temperature zone. It is hence assumed in this study that the temperature is uniform at a prescribed temperature in the uniform temperature zone to determine the value of rate constant at the prescribed temperature.

To eliminate the influence of conversion in the variable temperature zone, the temperature distribution in the variable temperature zone was fixed. In addition, the total gas flow rate was maintained unchanged and the residence time of reactant gas mixture in the uniform temperature zone was varied by different locations of reactant-feeder outlets. Under such conditions, the extents of conversion, X_{A1} and X_{A2}, corresponding to two different residence times, τ_1 and τ_2, are correlated, with subscript A as SiO, to determine rate constant k by[7]

$$\ln\frac{1-X_{A1}}{1-X_{A2}} = k\,(\tau_2-\tau_1) \tag{1}$$

Figure 3 shows an Arrhenius plot of rate constants, k, so obtained at three different temperatures. An apparent activation energy is evaluated from this plot to be 180 kJ/mol. Consequently, the rate equation for the ammonolysis of SiO may be represented in the range investigated, when NH_3 is in large excess over SiO, as

$$-r_A = 1.64\times10^6\ e^{-180\times10^3/RT}\ C_{SiO} \quad [\text{mol/m}^3\cdot\text{sec}] \tag{2}$$

where C_{SiO} is the concentration of SiO in the units of mol/m^3.

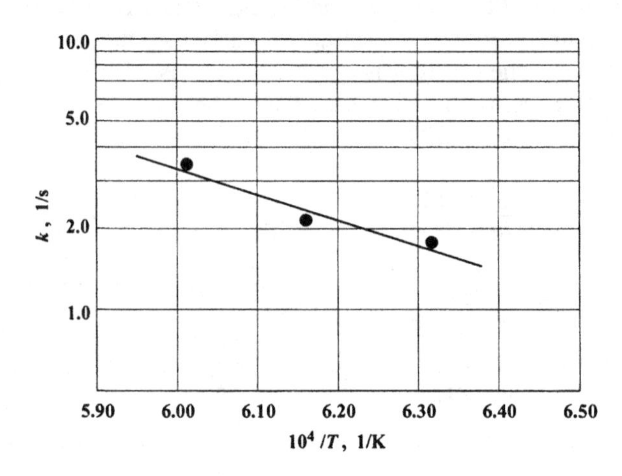

Figure 3. Arrhenius plot of pseudo first order rate constant *k*.

Production Rate of Whiskers

There are two specific locations where whiskers form: (1) at the exit of SiO generator and (2) on the reactor wall, region R1, as described earlier. It is suspected that an involvement of solid surface is essential for whisker formation. However, it has not been clearly understood why whiskers only form at these specific locations. To better understand the performance of reactor, this research has used a mathematical model to predict correlation between whisker formation and gas flow as well as reactant gas concentrations in the reactor.

Mathematical model - Simplification of reactor geometries: To simplify the reactor geometries, rectangular coordinates are selected for simulating velocity and concentration profiles at the plane containing all the three axes of reactor, SiO generator, and NH_3 feeder. One of the coordinates, the *z*-axis in this case, was taken as that extending infinitely in the direction perpendicular to the plane.

There are a few more assumptions that have been made in this study: (1) the system is isothermal and isobaric at steady state, (2) the gas-mixture is a constant-density, Newtonian fluid, and the flow is laminar, and (3) SiO and NH_3 are diluted in argon as a background fluid. Based on these assumptions, the equations of change were solved based on the finite volume method using a computational fluid dynamic modeling software (STAR-CD v. 3.102A). To avoid numerical instabilities, under-relaxation was used in iterative calculations as a weighted mean of the previous iteration value and the current one. The detailed computational algorithms should be referred to elsewhere.[8]

It should be noted that, due to the limitations associated with the software, the reaction between SiO and NH_3 as well as the dissociation of NH_3 could not be incorporated in the model simulation. Because the concentrations of these two species in the gas phase are very low, the velocity profiles obtained by the simulation may not have been affected by ignoring these two reactions. However, the concentration profiles may not have been well predicted by the simulation, except near the feeder outlets of these two reactants, where neither the ammonolysis of SiO nor NH_3 dissociation is significant.

Velocity profiles in the reactor: Circulation flow has been identified at two locations: (1) at the end of pipes between two feeder exits and (2) on the reactor wall next to the SiO generator outlet. These two locations are exactly the same as those where whiskers form when the same operating conditions are used in the actual reaction runs. Table III indicates agreements between the locations where whiskers form and those where stagnation appears in the simulation. It has been proved by many case studies by simulation and experimental runs that whiskers form at locations where stagnation of flow occurs right next to solid surface.

Table III. Correlation between locations of whisker formation and stagnation formation.

Location of stagnation	0	0.5	1	6	8.1
Location of whisker formation	0-1.5	1	3	7	8.5

Rate expression for whisker formation: It has been assumed that the simulated concentration profiles around the SiO generator outlet may not be very different from those in the reactor. Based on this assumption, the concentration of NH_3 and that of SiO predicted by the simulation were used for developing a rate expression for the whisker formation at the SiO generator. The formation rates of whiskers were determined by the mass of whiskers collected divided by the area where the whiskers were collected and by the time for reactor operation. The concentration of NH_3 as well as that of SiO was evaluated as an averaged value over the entire area in which whiskers formed.

Because it has already been shown that the rate of silicon nitride powder formation is of first order with respect to SiO concentration, it has been assumed that the whisker formation rate has the same SiO concentration dependency. The whisker formation rates at 1350°C may then be represented by

$$-r_{whiskers} = 0.158\, C_{SiO}\, C_{NH_3}{}^{0.31} \qquad \text{mol/s} \cdot \text{m}^2 \tag{3}$$

where C_{NH_3} is the concentration of NH_3 in the units of mol/m^3. The low exponent of about 1/3 to the NH_3 concentration implies multiple reaction pathways in the formation of whiskers, involving active intermediates, such as NH and NH_2, resulting from the NH_3 dissociation.[9] However, because the primary goal of this study was the synthesis of nano-sized silicon nitride powder, no further study was conducted for the whisker formation mechanism.

CONCLUSIONS

A vertical tubular-flow reactor was built to study the synthesis of nano-sized silicon nitride powder via the ammonolysis of SiO vapor at temperatures ranging from 1300°C to 1400°C. The effects of feed concentrations of NH_3 and SiO on the kinetics of nano-sized powder synthesis were investigated experimentally. The rate of nano-sized powder synthesis was independent of NH_3 concentration when NH_3 was in large excess over SiO, while the extent of SiO conversion was independent of its feed concentrations. Based on these findings, a rate expression of first order with respect to SiO concentration was proposed, and the magnitude as well as temperature dependency of apparent first-order rate constant was determined. The flow of gas in the reactor was analyzed using a computational flow dynamic program. The locations of whisker formation were correlated to the locations of gas-flow stagnation. Based on predicted reactant-gas concentrations, a power law rate expression for whisker formation was proposed, having implied a complex surface mechanism for whisker formation.

REFERENCES

[1]A. M. Sage and J H. Histed, "Applications of Silicon Nitride," *Metallurgy*, No. 8, 196-212 (1961).

[2]H. Gleiter, "Nanocrystalline Materials," *Prog. Mater. Sci.*, **33** [4] 223-315 (1989).

[3]D. Lin & S. Kimura, "Kinetics of Silicon Monoxide Ammonolysis for Nanophase Silicon Nitride Synthesis," *J. Am. Ceram. Soc.*, **79** [11] 2947-55 (1996).

[4]J. Szepvolgyi, I. Mohai, and J. Gubicza, "Atmospheric Ageing of Nanosized Silicon Nitride Powders," *J. Mater. Chem.*, **11**, 859-63 (2001).

[5]R. V. Weeren, E. A. Leone, S. Curran, L. C. Klein, and S. C. Danforth, "Synthesis and Characterization of Amorphous Si_2N_2O," *J. Am. Ceram. Soc.*, **77** [10] 2699-702 (1994).

[6]H. Wada, M. Wang, and T. Tien, "Stability of Phases in the Si-C-N-O System," *J. Am. Ceram. Soc.*, **71** [10] 837-40 (1998).

[7]O. Levenspiel, "Ideal Reactor for a Single Reaction"; pp. 102-103 in *Chemical Reaction Engineering*, 3'rd ed., John Wiley & Sons, New York, 1999.

[8]J. D. Anderson, Jr., "Computational Fluid Dynamics: The Basic with Applications," Int. ed., McGraw-Hill, 1995.

[9]A. B. F. Duncan and D. A. Wilson, "Intermediate Products in the Thermal Decomposition of Ammonia," *J. Am. Chem. Soc.*, **54** [1] 401-402 (1932).

AMBIENT CONDITION SYNTHESIS AND CHARACTERIZATION OF NANOCRYSTALLINE $BATIO_3$

Xinyu Wang and Burtrand I. Lee
School of Materials Science and Engineering
Clemson University, SC 29634

Michael Hu,* E. Andrew Payzant, and Douglas A. Blom
Oak Ridge National Laboratory
Oak Ridge, TN 37831
**Corresponding author*: huml@ornl.gov, 865-574-8782

ABSTRACT

Nanocrystalline $BaTiO_3$ particles have been prepared by the ambient condition sol (ACS) process that involves chemical precipitation of a titania gel and heated reflux processing of the gel in a Ba^{2+} containing solution. Among the processing variables, higher initial pH and Ba/Ti molar ratio led to smaller crystallite size of $BaTiO_3$ powders. The morphology of the $BaTiO_3$ powders was affected by the anion groups in the precursor species. Organic mineralizer, tetramethylammonium hydroxide (TMAH), can adsorb on the $BaTiO_3$ nuclei and further inhibit particle growth. Addition of a polymer dispersant during $BaTiO_3$ synthesis led to a smaller particle size and increased redispersibility of the particles in water.

INTRODUCTION

Barium titanate ($BaTiO_3$) based ceramics have been widely used in multilayer ceramic capacitors (MLCC), transducers, positive temperature coefficient of resistance (PTCR) thermistors, etc., due to their excellent dielectric and ferroelectric properties [1]. Traditionally, $BaTiO_3$ is prepared by solid-state reaction of fine barium carbonate ($BaCO_3$) and titanium dioxide (TiO_2) powders at high temperatures around 1000 °C followed by grinding [2, 3]. This method, however, leads to large $BaTiO_3$ particles (usually above 1 μm) with a wide size distribution and uncontrolled, irregular morphologies. All of these can lead to poor electrical properties and reproducibility of sintered ceramics. As a result of the use of relatively impure starting materials and ball-milling operations, impurities such as Si, Al, S, P, and carbonates are usually introduced in conventionally prepared ceramics [4, 5].

The MLCC industry is continuing intensive efforts to reduce component size by decreasing the dielectric layer thickness down to 1-2 μm [6]. Thus, $BaTiO_3$ nanoparticles have great advantages over micrometer sized $BaTiO_3$ particles in enhancing film homogeneity and reducing microstructural defects when a single ceramic layer is less than ~1 μm. In the past decades, extensive studies have been conducted to produce nanosized $BaTiO_3$ powders with narrow particle size distribution, controlled morphology, and high purity. $BaTiO_3$ nanocrystals have been synthesized by various techniques: hydrothermal method [7-26], sol-gel process [27-32], low temperature aqueous synthesis (LTAS) [33-35], low temperature direct synthesis (LTDS) [36-37], combustion synthesis [38], oxalate coprecipitation route [39-40], microwave heating [41], and micro-emulsion process [42]. Among these methods, wet chemical synthesis, such as hydrothermal and LTAS, involve reactions in aqueous media under strong alkaline conditions. These provide a promising way to produce nanocrystalline $BaTiO_3$ with narrow particle size distribution, high purity, and controlled morphology of resulting particles at a low temperature and under a mild pressure.

Hennings et al. [9] synthesized $BaTiO_3$ powders from Ba-Ti acetate gel precursors by a two-step route: transparent gels of Ba-Ti acetate was formed by sol-gel process followed by hydrothermal treatment. Their thermaogravimetric analysis (TGA) and titration results revealed stoichiometric composition of Ba-Ti gels. They also observed by transmission electron microscopy (TEM) that nucleation began after 30 min: large number of tiny (<5 nm) $BaTiO_3$ nuclei formed on the surface of the gel. The reaction completed after 10 hours and 200-300 nm $BaTiO_3$ particles were produced. Xia et al. [12] studied the effect of the reaction temperature, Ba/Ti ratio, type of precursors on the phase composition, the size and the morphology of the $BaTiO_3$ powders under a hydrothermal condition. They found that the higher the temperature, the basicity, and the Ba/Ti ratio, the easier to form $BaTiO_3$ nanocrystalline. They claimed that tetragonal $BaTiO_3$ was obtained at Ba/Ti ratio above 2:1.

As for the morphology control, Lu et al. [22] reported synthesizing $BaTiO_3$ by hydrothermal method in the presence of a surface modifier (Tween 80). Ultrafine (< 100 nm) $BaTiO_3$ particles with narrow size distribution were obtained. Those powders also showed good re-dispersibility. Hu and co-workers [24] produced $BaTiO_3$ powders in a two-stage approach, which involves dielectric tuning solution (DTS) precipitation of near-monodispersed TiO_2 spheres in a mixed solvent of isopropanol and water, followed by low temperature hydrothermal conversion of TiO_2 spheres in $Ba(OH)_2$ solutions. The resulting $BaTiO_3$ microspheres were uniform and well dispersed.

Wada et al. [36] established LTDS method to synthesize nanosized $BaTiO_3$ powders. They proposed that the heat of neutralization by mixing a strong acid and a base can be used as the driving force for the formation of $BaTiO_3$. Moreover, $BaTiO_3$ could be directly synthesized from Ba and Ti ions, not via intermediate. For the first time, $BaTiO_3$ particles with crystallite size less than 10 nm were reported.

However, since water is the common medium in the hydrothermal, LTAS and LTDS, it's inevitable to incorporate much OH species into $BaTiO_3$ lattice as unwanted defects [43]. Hennings et al. [9, 10, 17] characterized hydrothermal $BaTiO_3$ powders and found OH groups were incorporated in the perovskite lattice. A defect chemical model was derived and the lattice defects were calculated from TGA data. Clark and co-workers [20] had similar observations. It is well known that such defects can lead to deviation of stoichiometry, poor sintering density, inhomogeneous microstrucure of green tapes, and quality and reproducibility degradation of final products [17]. Therefore, reducing such kind of defects is challenging for $BaTiO_3$ synthesis.

To prepare $BaTiO_3$ crystallites with fewer defects, Wada et al. [37] modified the LTDS method, using highly concentrated aqueous solution of $Ba(OH)_2$. $BaTiO_3$ crytallites with sizes around 16 nm were obtained, however, hydroxyl group was still detected as an impurity. Moreover, Ba/Ti ratio was determined as 0.786 ± 0.005, indicating the presence of many Ba vacancies in the crystallies.

The newly developed solvothermal method, which involes using an organic solvent instead of water [44, 45], may provide an alternative to produce defect free $BaTiO_3$ nanoparticles. In Chen and Jiao's work [45], Ba and Ti alkoxides were mixed in $HOCH_2CH_2OCH_3$ and hydrolyzed to form a precursor gel. The gel was dried under vacuum and the dried xerogel was solvothermally treated in the alcohol at 140 - 240 °C for up to 144 hours. Cubic $BaTiO_3$ powders with small particle size of 20-60 nm and narrow particle size distribution were obtained. It was observed that solvothermal reaction was slower than hydrothermal, attributing to the low solubility of precursor gel in alcohol. However, no defect analysis was reported in their paper.

O'Brien et al. [46] developed an "injection-hydrolysis" procedure to produce high purity, crystalline nanoscale materials. Barium titanium ethyl hexano-isopropoxide was used to ensure the precise stoichiometry and injected into the mixture of diphenyl ether and oleic acid at 140 °C under argon. The mixture was cooled to 100 °C and hydrogen peroxide was injected and the mixture was maintained and stirred at 100 °C for over 48 h to promote further hydrolysis and crystallization. Monodisperse nanoparticles of $BaTiO_3$ with diameters ranging from 6 to 12 nm were obtained. However, no defect analysis was carried out in their research. Moreover, since hydrogen peroxide was used, existence of lattice hydroxyl groups in their $BaTiO_3$ nanoparticles is expected.

In our previous work [47, 48], we reported briefly the synthesis of nanocrystalline $BaTiO_3$ powders by a novel method—ambient condition sol (ACS) process. In this paper, the factors impacting the morphology and properties of $BaTiO_3$ particles from ACS process are presented.

EXPERIMENTAL

All the chemical reagents used in this research were analytical grade and no further purification was performed before use. The experimental procedure was described in our previous work [47, 48]. To study the influence of experimental parameters on the characteristics of final $BaTiO_3$ products, a series of $BaTiO_3$ samples have been prepared as described in Table 1.

Table 1. List of $BaTiO_3$ samples prepared by the ACS process

Sample ID	Ba/Ti	Mineralizer	pH	Reaction Time	Media
BT1	1.5:1	KOH	12.0	20h	H_2O
BT2	1.5:1	KOH	14.0	20h	H_2O
BT3	1.5:1	KOH	14.2	20h	H_2O
BT4	1.5:1	KOH	14.0	6h	H_2O
BT5	3:1	KOH	14.0	20h	H_2O
BT6*	1.5:1	KOH	14.0	20h	H_2O
BT7**	1.5:1	KOH	14.0	20h	H_2O
BT8	1.5:1	TMAH	14.0	20h	H_2O
BT9***	1.5:1	KOH	14.0	20h	H_2O

* $Ti(OC_2H_5)_4$ as Ti source in precursor
** $Ba(OOCCH_3)_2$ as Ba source in precursor
*** With 2 mg APA per ml.

Dynamic light scattering (DLS) technique was applied to estimate the particle size and particle size distribution of resulting $BaTiO_3$ powders. The samples were diluted with deionized water and ultrasonicated for 15 min before analysis. A low power (10 mW) He-Ne laser tube generated an incidental beam (wavelength 632.8 nm) going through the sample solution. The light that was scattered from the particles in solution was monitored by a photomultiplier tube that was placed at a 90° angle to the incident beam. The data processing was performed using a digital correlator (Model BI-9000AT, Brookhaven Instruments Corp., Holtsville, NY), which collected pulsed signals from the photomultiplier and generated an autocorrelation function.

Room temperature X-ray diffraction (RTXRD, Scintag PAD V using CuK_α with λ=0.15406 nm, 2θ = 20-80° with a scan rate of 1°/min) was used for crystalline phase identification of the particle samples and determination of crystallite size in the powders. The crystallite size was calculated by the Scherrer Eqn:

$$d_x = \frac{0.94\lambda}{\beta \cos\theta} \quad [1]$$

where d_x is the crystallite size, λ is the wave length of X-ray, β is the full-width at half-maximum (FWHM), θ is the diffraction angle. The (200) peak was used to calculate the crystallite size.

The morphology of $BaTiO_3$ powders was examined by scanning electron microscopy (SEM) and TEM. A small amount of $BaTiO_3$ powders were pressed on a carbon tape, which attached to the flat surface of brass sample stub. Air spray was used to blow out the $BaTiO_3$ powders that were not adhered to the carbon tape. The $BaTiO_3$ powders were then covered by a gold coating via plasma sputtering (Hammer 6.2 Sputtering System, Anatech) to create a conductive surface layer that was necessary for SEM imaging. To prepare TEM samples, a tiny amount of $BaTiO_3$ particles was dispersed into iso-propanol by grinding in an agate mortar. A copper grid with a supported thin carbon film was dipped into the suspension, removed and dried on a filter paper.

RESULTS AND DISCUSSION

1. Effect of Solution pH

To study the impact of the solution pH used for reflux conversion of precipitated titania gel, $BaTiO_3$ samples were prepared at pH 12, 14, and 14.2 (corresponding to samples BT1, BT2, and BT3). The RTXRD patterns in Figure 1 indicate that $BaTiO_3$ samples prepared at pH 12 (BT1) was amorphous, while the $BaTiO_3$ powders prepared above pH14 (samples BT2 and BT3) were well crystallized, cubic phase. It can be concluded that the alkalinity plays an important role in the crystallization of $BaTiO_3$ under ACS conditions. According to the thermodynamic model [49] for the hydrothermal synthesis, phase-pure $BaTiO_3$ could only be obtained at a pH higher than 13.5. Moreover, the diffraction peaks of BT3 are slightly broader than BT2, indicating existence of smaller nanosized crystallites in the BT3 powder. Calculated from the full-width at half-maximum (FWHM) of {200} peak by the Scherrer equation, the crystallite size of BT3 is 43 ± 2 nm, which is indeed smaller than that of BT2 with 52 ± 2 nm (Table 2). At the higher pH, it is believed that greater repulsive forces between the $BaTiO_3$ nuclei inhibit further growth.

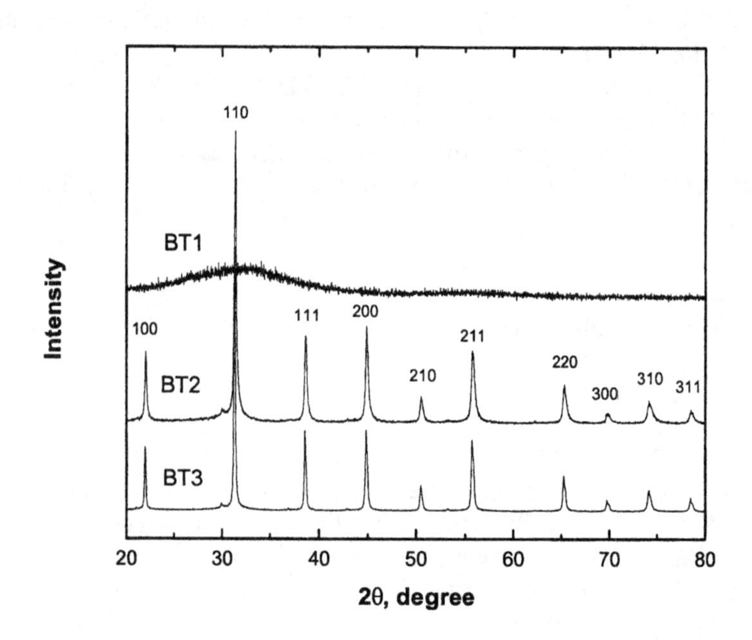

Figure 1. RTXRD patterns of $BaTiO_3$ samples prepared at various values of pH: pH 12 (BT1), 14 (BT2) and 14.2 (BT3).

Table 2. Crystallite size and particle size of $BaTiO_3$ powders by ACS synthesis.

	BT2	BT3	BT4	BT5	BT6	BT7	BT8	BT9
Crystallite size, nm	52±2	43±2	50±2	42±2	50±2	53±2	<10	47±1
Particle size, nm	126±15	91±18	112±20	79±10	207±23	164±38	146±16	82± 13

Figure 2 shows the micrographs of $BaTiO_3$ powders synthesized under different pH and Ba/Ti ratio. No individual particles or crystallites could be seen in the chalky mass of BT1 sample gelled at pH 12 (Figure 2 (a)), which agrees with the RTXRD pattern in Figure 1. BT2 particles formed at pH 14 are nearly spherical with slight agglomeration and narrow particle size distribution. DLS results show that the mean particle size is 126 ± 15 nm (Table 2, second row). Comparing with BT2, BT3 particles formed at pH 14.2 are smaller in particle size but slightly more agglomerated. DLS data reveal that the mean particle size of BT3 is 91 ± 18 nm. The sizes of spherical shaped particles as shown in the SEM images (Figure 2) are nearly agreeable with the size measured by the DLS. This indicates that the spherical particles are close to fully dispersed in water and the somewhat agglomerated state shown in Figure 2 images may simply due to artifacts of sample drying on SEM stub surface.

The above results show a trend that a higher pH of reaction media leads to smaller particle size of $BaTiO_3$ for the reason given above for the crystallite size.

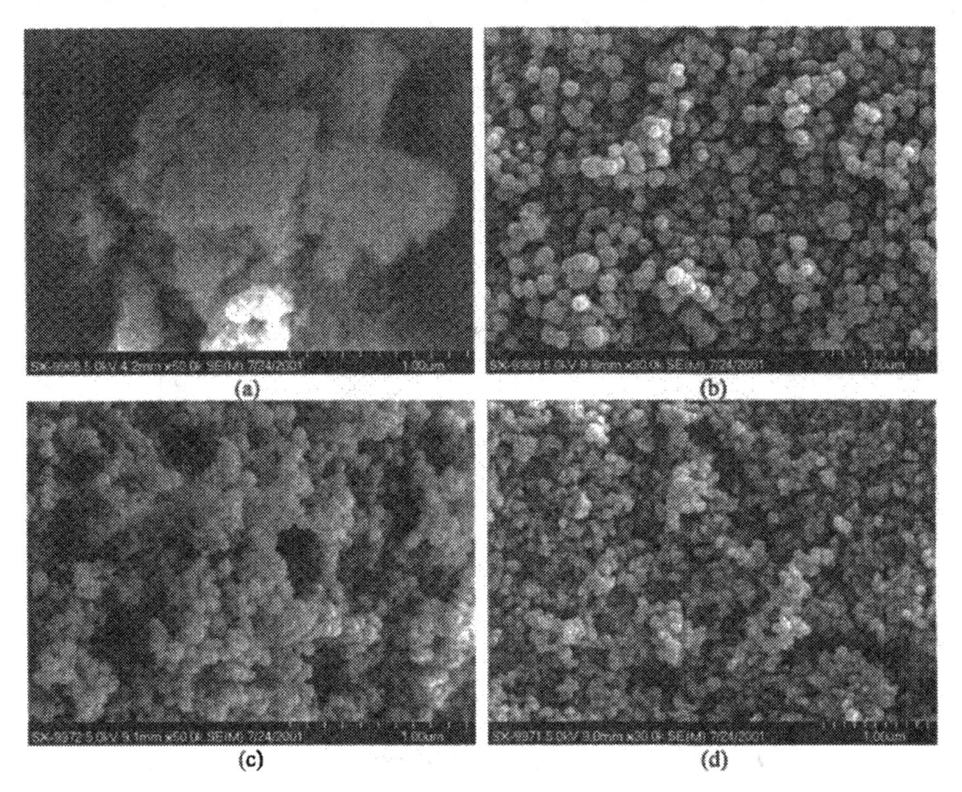

Figure 2. SEM micrographs of $BaTiO_3$ powders. (a) BT1 (pH 12, Ba/Ti=1.5); (b) BT2 (pH 14, Ba/Ti=1.5); (c) BT3 (pH 14.2, Ba/Ti=1.5); (d) BT5 (pH 14, Ba/Ti=3).

2. Effect of Time for Reflux Conversion of Precursor Titania Gels

BT4 was prepared at pH 14.0 with Ba/Ti=1.5 in aqueous media for 6 hours. The RTXRD pattern reveals well crystallized cubic phase (Figure 3). The crystallite size of BT4 calculated from FWHM of {200} peak by the Scherrer equation is 50 ± 2 nm. DLS and SEM results indicate that the morphology and particle size of BT4 are similar to that of BT2 which was prepared under the same condition but for 20 hours (particle size of BT4 by DLS measurement is 112 ± 20 nm). Our previous study [48] showed that the formation of $BaTiO_3$ nanoparticles followed an "in-situ transformation" mechanism and was a rapid process (within 30 min). Once the particle is formed, the effect of extending the reaction time on the morphology, particle size, crystallite size, and crystallinity of $BaTiO_3$ particles is trivial.

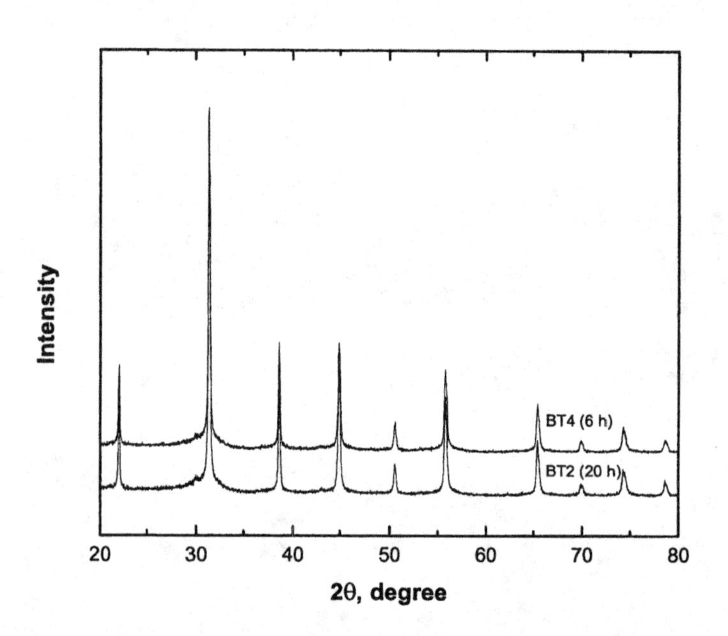

Figure 3. RTXRD of $BaTiO_3$ powders prepared with different reaction time.

3. Effect of the Initial Ba/Ti Ratio

To study the influence of Ba/Ti ratio of starting materials under the same pH, reaction temperature and time, $BaTiO_3$ powders (BT5) was prepared with initial Ba/Ti=3:1. No split of {200} peak at 2θ = 44.95° is found in RTXRD pattern (Figure 4), indicative of cubic phase structure with symmetry Pm3m. Calculating from the FWHM of {200} peak by Scherrer's formula, the crystallize size of BT5 sample is 42 ± 2 nm, slightly smaller than that of BT2.

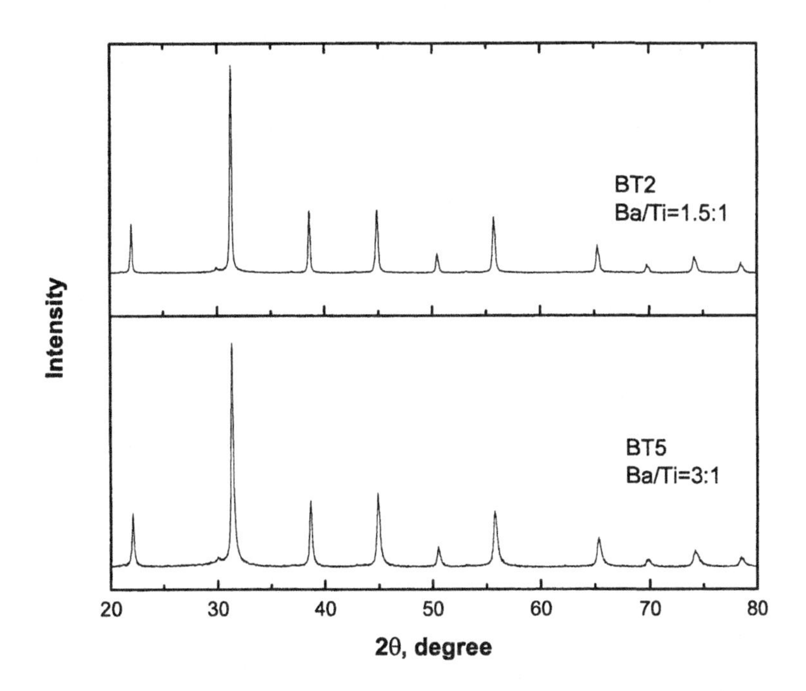

Figure 4. RTXRD patterns of $BaTiO_3$ powders prepared with different initial Ba/Ti ratio.

Figure 2(d) represents the morphology of BT5 powders. Comparing with BT2 powders (Ba/Ti=1.5), noticeably smaller particles with smaller particle size distribution can be found in BT5 sample (Ba/Ti=3). One can also notice the increased degree of agglomeration of BT5 particles. The DLS results show that the mean particle size of BT5 is 79 ± 10 nm. Wada et al. [36] also reported similar observation for $BaTiO_3$ powders prepared by LTDS method. At a fixed reaction temperature, increasing Ba/Ti ratio resulted in smaller crystallite size (37.0 nm when Ba/Ti=5 and 12.9 nm when Ba/Ti>35).

4. Influence of Anions

To study the influence of the chemical precipitate precursor on the morphology of ACS $BaTiO_3$, two different $BaTiO_3$ samples were prepared. BT6 was prepared from $Ti(OC_2H_5)_4$ and $BaCl_2$, while BT7 was synthesized from $TiCl_4$ and $Ba(OOCCH_3)_2$. All the experimental parameters, such as temperature, time,

Ba/Ti ratio and pH were the same as those of BT2. No peak split of {200} and {002} at around $2\theta = 44.95°$ were found in RTXRD pattern of BT6 and BT7 (Figure 5), indicating a cubic phase. The crystallite sizes of BT6 and BT7 are 50 ± 2 nm and 53 ± 2 nm, respectively, similar to BT2.

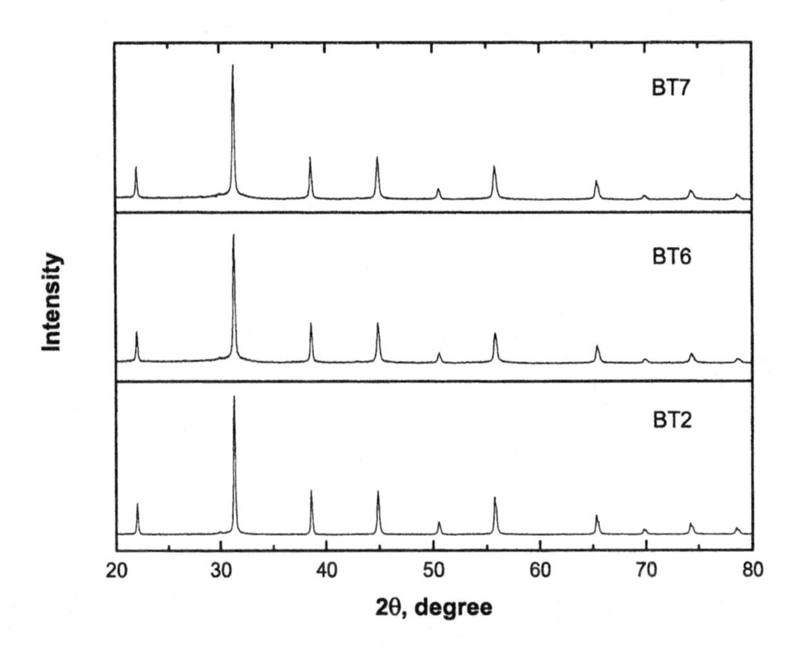

Figure 5. Comparison of RTXRD patterns of BT2 ($BaCl_2$ and $TiCl_4$), BT6 ($BaCl_2$ and $Ti(OC_2H_5)_4$), and BT7 ($BaAc_2$ and $TiCl_4$) powders

The morphology of BT6 and BT7 shown in Figure 6 is different from the morphology of BT2 shown in Figure 2(b). Both BT6 and BT7 particles are spherical. However, it can be seen that there is more particle agglomeration in BT6 than in BT7 sample which exhibits a broad particle size distribution from 50 to 200 nm. DLS data show that the mean particle size of BT6 is 207 ± 23 nm, and 164 ± 38 nm for BT7.

Based on the above results, it can be concluded that the morphology of $BaTiO_3$ particles is affected by the types of anions in the initial chemical precursors. In comparing the precursors for BT6 and BT7, BT7 has more ionic strength than BT6, and the greater ionicity and coordination role of the anions may result in the smaller particle size and better dispersion of BT7. Similar results have also been reported by other researchers [12, 50].

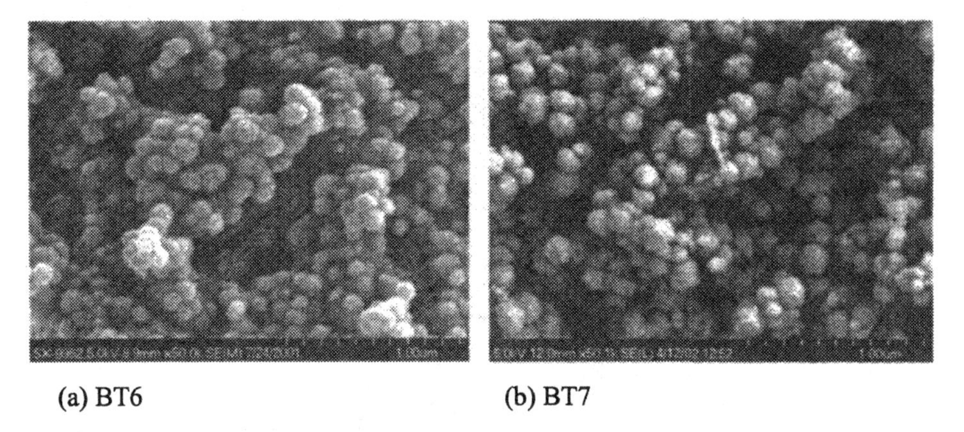

(a) BT6 (b) BT7

Figure 6. SEM micrographs of BT6 ($BaCl_2$ and $Ti(OC_2H_5)_4$) and BT7 ($BaAc_2$ and $TiCl_4$) powders prepared by ACS process.

5. Effect of Mineralizer

The titania precipitate precursors for ACS process were obtained by chemically precipitating starting materials in a basic solution. KOH may be replaced by other strong base, and in turn some property changes of the final products could be expected. In this research, tetramethylammonium hydroxide (TMAH) was used, instead of KOH, to adjust pH of the precursor solution to 14.0 and the resulting $BaTiO_3$ was labeled as BT8. RTXRD pattern (Figure 7) of the BT8 powders shows cubic phase, similar to the other previous results. However, significant peak broadening was observed, indicating either poor crystallinity or existence of very small crystallite size (around 10 nm) in the sample.

All BT8 particles are spherical with a soft appearance (Figure 8(a)) as compared to those particles prepared with KOH. Moreover, each individual BT8 particle is found to comprise of large amounts of much smaller primary particles at a closer observation. The TEM micrograph of BT8 (Figure 8(b)) also indicates that each particle is an aggregate of many small crystals. Thus the broad RTXRD pattern in Figure 7 must be caused by the crystallite size less than 10 nm.

Similar to KOH as a strong base, TMAH undergoes complete dissociation in water:

$$N(CH_3)_4^{+}OH^{-} \longrightarrow N(CH_3)_4^{+} + OH^{-} \qquad [2]$$

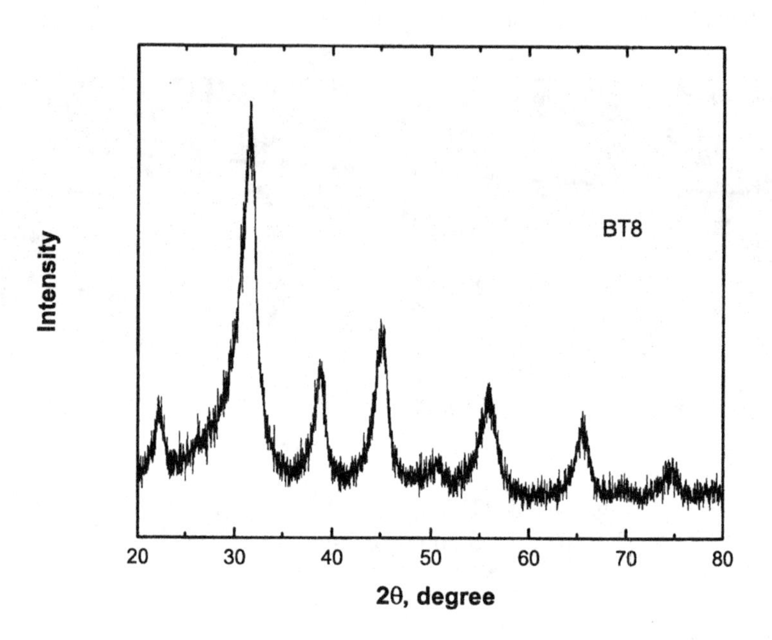

Figure 7. RTXRD pattern of BT8 (with TMAH) powder shows significant peak broadening.

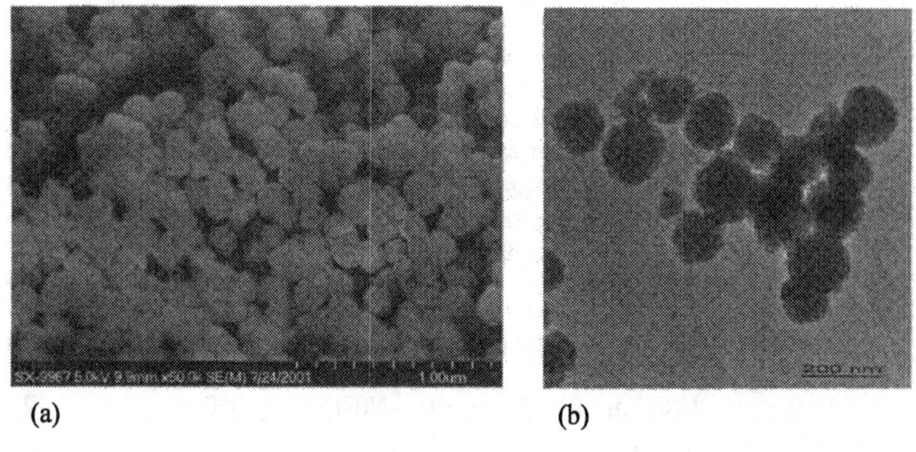

Figure 8. SEM (a) and TEM (b) mirographs of BT8 powders.

While the precursor gels are forming via chemical precipitation, some cations are adsorbed onto the cluster of precursor gels. Obviously $N(CH_3)_4^+$ ion is much larger than K^+ and can stabilize and inhibit further growth of the nuclei of barium titanate crystals. However, since the primary particles/crystals are small (<10 nm by XRD), they tend to aggregate into large particles. The high weight loss (13.2 wt%) at 800 °C in TGA curve (not shown in this paper) confirms that there is large amount of organic species and water in BT8 powders. DLS results show the particle size of BT8 is 146 ±16 nm, indicating the aggregates of tiny crystals are difficult to break.

6. Effect of Polymer Additive

A $BaTiO_3$ sample labeled as BT9 has been made in the presence of ammonium polyacrylate (APA) with a mean molecular weight of 6000 g/mol at a concentration of 0.002 g/ml. All other experimental parameters, such as the pH, Ba/Ti, and reaction time were similar to BT2.

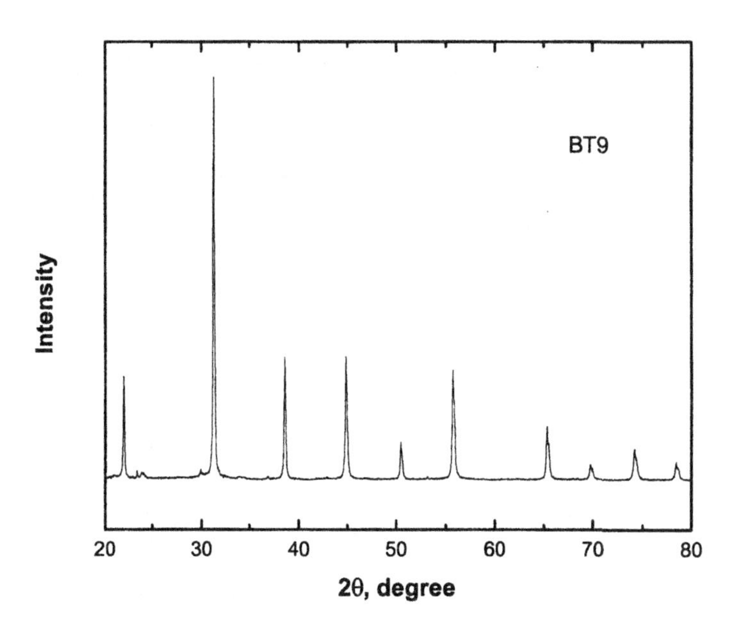

Figure 9. RTXRD pattern of BT9 powders.

Figure 10 SEM micrograph of BT9 powders prepared with APA.

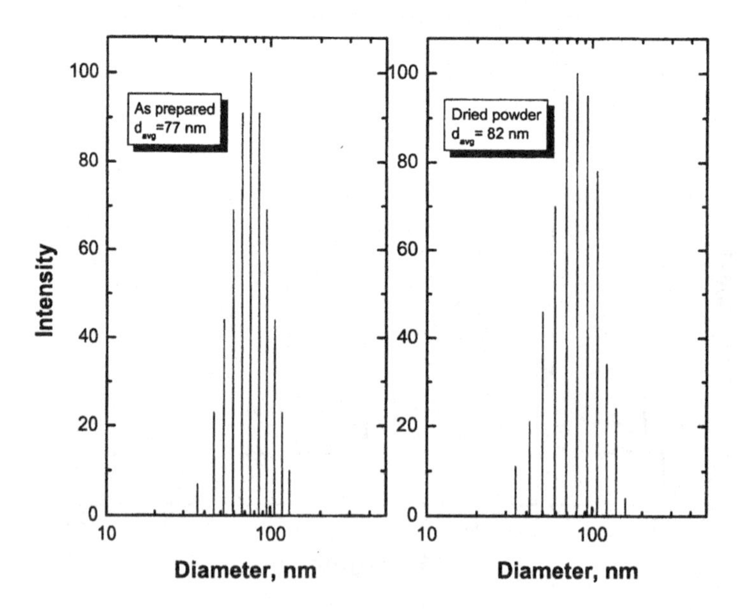

Figure 11. Particle size distribution of BT9, as-prepared (left) and re-dispersed after air drying (right).

Figure 9 shows the RTXRD pattern of BT9 powders, which represents pure cubic phase. The crystallite size of BT9 calculated by the FWHM of {200} peak is 47 ± 1 nm. Near spherical $BaTiO_3$ powders are observed in the SEM micrograph (Figure 10). Moreover, BT9 particles are smaller than those of BT2, indicating that the particle growth is inhibited by the adsorbed APA molecules.

The DLS results (Figure 11) also indicate that the BT9 powders possess good re-dispersibility. The particle size of as-prepared (wet powder) and re-dispersed (dried powder) are measured by DLS technique. The results show that the particle size of re-dispersed BT9 powders is almost the same as the BT9 particles before drying, indicating that introducing polymer dispersant during particle synthesis can result in re-dispersible powders.

CONCLUSIONS

This paper reports an ambient condition sol (ACS) process for producing nanocrystalline $BaTiO_3$ particles. The ACS process involves chemical precipitation of a titania gel and heated reflux processing of the gel in a Ba^{2+}-containing aqueous solution. Among the processing variables, higher initial pH and Ba/Ti molar ratio led to smaller crystallite size of $BaTiO_3$ powders. Organic mineralizer, tetramethylammonium hydroxide (TMAH), can adsorb on the $BaTiO_3$ nuclei and further inhibit primary particle/crystal growth. The morphology of the $BaTiO_3$ powders was affected by the anion groups in the precursor species. Addition of a polymer dispersant (such as APA) during $BaTiO_3$ synthesis led to a smaller particle size and increased redispersibility of the particles in water.

ACKNOWLEDGEMENTS

This work is sponsored by the Division of Materials Sciences, Office of Basic Energy Sciences, of the U.S. Department of Energy, and partially sponsored by National Science Foundation with Grant Number DMR-9731769. Research is also sponsored in part by the Assistant Secretary for Energy Efficiency and Renewable Energy, Office of FreedomCAR and Vehicle Technologies, as part of the High Temperature Materials Laboratory User Program, Oak Ridge National Laboratory, managed by UT-Battelle, LLC, for the U.S. Dept. of Energy under contract DE-AC05-00OR22725.

References

1. J. Nowotny and M. Rekas, “Electronic Ceramic Materials”, pp1-144, Trans Tech, Zurich, Switzerland, 1992.
2. A. Bauger, J. C. Mutin, and J. C. Niepce, *J. Mater. Sci.*, **18** 3041-3046 (1983).
3. M. S. H. Chu and A. W. I. M. Rae, *Am. Ceram. Soc. Bull.*, **74** 69-72 (1995).
4. P. P. Phule and S. H. Risbud, *J. Mater. Sci.*, **25** 1169-1183 (1990).

5. M. C. B. Lopez, G. Fourlaris, B. Rand, and F. L. Riley, **J. Am. Ceram. Soc.**, **82** 1777-1786 (1999).
6. J. M. Wilson, *Am. Ceram. Soc. Bull.*, **74** 106-110 (1995).
7. W. Hertl, *J. Am. Ceram. Soc.*, **71** 879-883 (1988).
8. R. Vivekanandan and T. R. N. Kutty, *Powder Technology*, **57** 181-192 (1989).
9. D. Hennings, G. Rosenstein and H. Schreinemacher, *J. Euro. Ceram. Soc.*, **8** 107-115 (1991).
10. D Hennings and S. Schreinemacher, *J. Euro. Ceram. Soc.*, **9** 41-46 (1992).
11. P. K. Dutta, R. Asiaie, S. A. Akbar, and W. D. Zhu, *Chem. Mater.*, **6** 1542-1548 (1994).
12. C. T. Xia, E. W. Shi, W. Z. Zhong, and J. K. Guo, *J. Euro. Ceram. Soc.*, **15** 1171-1176 (1995).
13. J. O. Eckert Jr., C. C. H. Houston, B. L. Gersten, M. M. Lencka and R. E. Riman, *J. Am. Ceram. Soc.*, **79** 2929-2939 (1996).
14. R. Asiaie, W. D. Zhu, S. A. Akbar, and P. K. Dutta, *Chem. Mater.*, **8** 226-234 (1996).
15. E. W. Shi, C. T. Xia, W. Z. Zhong, B. G. Wang, and C. D. Feng, *J. Am. Ceram. Soc.*, **80** 1567-1572 (1997).
16. W. D. Zhu, S. A. Akbar, R. Asiaie, and P. K. Dutta, *Jpn. J. Appl. Phys.*, **36** 214-221 (1997).
17. D. Hennings, C. Metzmacher and B. S. Schreinemacher, *J. Am. Ceram. Soc.*, **84** 179-182 (2001).
18. S. Urek and M. Drofenik, *J. Euro. Ceram. Soc.*, **18** 279-286 (1998).
19. P. Pinceloup, C. Courtois, A. Leriche, and B. Thierry, *J. Am. Ceram. Soc.*, **82** 3049-3056 (1999).
20. I. J. Clark, T. Takeuchi, N. Ohtori, and D. Sinclair, *J. Mater. Chem.*, **9** 83-91 (1999).
21. I. MacLaren and C. B. Ponton, *J. Euro. Ceram. Soc.*, **20** 1267-1275 (2000).
22. S. W. Lu. B. I. Lee, Z. L. Wang, and W. D. Samuels, *J. Crystal Growth*, **219** 269-276 (2000).
23. M. Z. Hu, G. A. Miller, E. A. Payzant, and C. J. Rawn, *J. Mater. Sci.*, **35** 2927-2936 (2000).
24. M. Z. Hu, V. Kurian, E. A. Payzant, C. J. Rawn, and R. D. Hunt, *Powder Technology*, **110** 2-14 (2000).
25. H. R. Xu, L. Gao and J. K. Guo, *J. Euro. Ceram. Soc.*, **22** 1163-1170 (2002).
26. E. Ciftci, M. N. Rahaman and M. Shumsky, *J. Mater. Sci.*, **36** 4875-4882 (2001).
27. H. Shimooka and M. Kuwabara, *J. Am. Ceram. Soc.*, **79** 2983-2985 (1996).
28. M. H. Frey and D. H. Payne, *Chem. Mater.*, **7** 123-129 (1995).
29. H. Matsuda, M. Kuwabara, K. Yamada, H. Shimooka and S. Takahashi, *J. Am. Ceram. Soc.*, **81** 3010-3012 (1998).

30. B. I. Lee and J. P. Zhang, *Thin Solid Films*, **388** 107-113 (2001).
31. H. Matsuda, N. Kobayashi, T. Kobayashi, K. Miyazawa and M. Kuwabara, *J. Non-Cryst. Solids*, **271** 162-166 (2000).
32. H. P. Beck, W. Eiser and R. Haberkorn, *J. Euro. Ceramic Soc.*, **21** 687-693 (2001).
33. P. Nanni, M. Leoni, V. Buscaglia, and G. Alipandi, *J. Euro. Ceram. Soc.*, **14** 85-90 (1996).
34. M. Leoni, M. Viviani, P. Nanni, and V Buscaglia, *J. Mater. Sci. Lett.*, **15** 1302-1304 (1996).
35. M. Viviani, J. Lemaitre, M. T. Buscaglia and P. Nanni, *J. Euro. Ceram. Soc.*, **20** 315-320 (2000).
36. S. Wada, H. Chikamori, T. Noma, and T. Suzuki, *J. Mater. Sci.*, **35** 4857-4863 (2000).
37. S. Wada, T. Tsurumi, H. Chikamori, T. Noma, and T. Suzuki, *J. of Crystal Growth*, **229** 433-439 (2001).
38. T. V. Anuradha, S. Ranganathan, T. Mimani and K. C. Patil, *Scripta Mater.*, **44** 2237-2241 (2001).
39. M. Stockenhuber, H. Mayer, and J. A. Lercher, *J. Am. Ceram. Soc.*, **76** 1185-1190 (1993).
40. P. K Gallagher and F. Schrey, *J. Am. Ceram. Soc.*, **46** 567-573 (1963).
41. Y. Ma, E. Vileno, S. L. Suib and P. K. Dutta, *Chem. Mater.*, **9** 3023-3031 (1997).
42. J. Wang, J. Fang, S. C. Ng, L. M. Gan, C. H. Chew, X. B. Wang, and Z. X. Shen, *J. Am. Ceram. Soc.*, **82** 873-881 (1999).
43. G. Busca, V. Buscaglia, M. Leoni, and P. Nanni, *Chem. Mater.*, **6** 955-961 (1994).
44. G. Demazeau, *J. Mater. Chem.*, **9** 15-18 (1999).
45. D. R. Chen and X. L. Jiao, *J. Am. Ceram. Soc.*, **83** 2637-2639 (2000).
46. S. O'Brien, L. Brus, and C. B. Murry, *J. Am. Ceram. Soc.*, **123** 12085-12086 (2001).
47. X. Wang, B. I. Lee, M. Hu, E. A. Payzant, D. A. Blom, *J. Mater. Sci. Lett.*, **22** 557-559 (2003).
48. X. Wang, B. I. Lee, M. Hu, E. A. Payzant, D. A. Blom, *J. Mater. Sci, Mater. in Elcetron.*, in press.
49. M. M. Lencka and R. E. Riman, *Chem. Mater.*, **5** 61-70 (1993).
50. M. Wu, R. Xu, S. H. Feng, *J. Mater. Sci.*, **31** 6201-6205 (1997).

FERROELECTRIC LITHOGRAPHY OF MULTICOMPONENT NANOSTRUCTURE

Xiaojun Lei, Sergei Kalinin, Zonghai Hu, Dawn A. Bonnell*
Department of Materials Science and Engineering
The University of Pennsylvania
3231 Walnut Street
Philadelphia PA 19104

ABSTRACT

Atomic polarization in PZT is manipulated to control surface electronic structure and local photochemical reactivity. Using photoreduction from aqueous solution, metal nanoparticles are produced in predefined locations on an oxide substrate. Through adjusting the photoreduction conditions, particle size and density can be varied. The photoreduction process is discussed with respect to wavelength, solution concentration, deposition time and reagents. Subsequently, organic molecules are reacted selectively to the particles. The process can be repeated to develop complex structures consisting of nanosized elements of semiconductors, metals, or functional organic molecules.

INTRODUCTION

Interesting local phenomena recently demonstrated include coulomb blockade[1], single-electron tunneling[2], surface enhanced Raman[3], and electron diffraction in a ring of iron atoms[4]. Phenomena such as these will play a critical role in designing future nanodevices that are expected to be smaller, faster, require less energy and produce less heat than current devices. To enable these phenomena in electronic, optical, and magnetic applications, much effort has been devoted to fabricating nanostructures. Two strategies are employed to produce structure at the nanoscale level. One, exemplified by traditional silicon technology, reduces the feature size by improving radiation source, photoresist, etc. Since the feature size scales from a whole wafer to a submicron line, this is referred to as "top-down" processing. In contrast, "bottom-top" approaches assemble nanoparticles, nanotubes and macromolecules through chemical interaction. The number of possible nanoscale building blocks is increasing rapidly. It is now possible to make molecular and nanometer scale elements with a variety of electrical, optical and chemical properties.[5-14] Semiconducting and metallic nanowires, organic molecules, carbon nanotubes and biological molecules have been designed to exhibit properties that might enable their application as field effect transistors, biochemical sensors or display devices.[15] It is generally accepted that self assembly in some form is required for nanofabrication, however, general self assembly produces 2D or 3D arrays, not complex devices. Some recent efforts have focused on self-assembling nanocomponents in a predefined area of a substrate.[16-21] Regions of organic monolayers (SAM) self-assembled on a substrate are modified, for example with atomic force microscopy (AFM) or e-beam lithography, and nanostructures are linked via the modified or unmodified monolayer to the substrate. While this class of processes shows promise, it is limited to a restricted set of materials.[22]

We report here a novel paradigm for the directed assembly of nanoscale elements that locates structures in predefined complex configurations. This process makes use of arrays of atomic polarization to direct site-specific chemical interactions. Polarization mediated assembly is based on the combination of two distinct phenomena: atomic polarization-dependent surface electronic structure in ferroelectric compounds and photo reactions. Polarization-dependent chemical reactivity of ferroelectric materials was demonstrated shortly after the discovery of ferroelectricity.[23] Perovskite titanates are extremely efficient catalysts in photooxidation and photoreduction processes. Recently, Rohrer *et. al.*[24] demonstrated domain specific photoreduction of aqueous metal cations on polycrystalline $BaTiO_3$. In this paper, reaction products are patterned onto PZT ($Pb(Zr_{0.2}Ti_{0.8})O_3$) thin films producing complex configurations of semiconductors, metallic nanoparticles and organic molecules. We refer to this process as ferroelectric nanolithography.

EXPERIMENTAL PROCEDURES

PZT thin films were prepared by a sol-gel method on $Pt/SiO_2/Si$ substrate. The film thickness was ~200nm and characteristic grain size was 50-100nm. Aqueous metal salt solutions of concentrations ranging from 10^{-6}M to 10^{-2}M were made from 99.9+% silver nitrate. A Hg/Xe arc lamp with a 300w power supply (66011, 68811, Oriel Instruments) and a monochromator (82-410, Jarrell Ash) were used for the radiation source. As a proof of concept, saturated hydrocarbons were reacted from 1×10^{-4} M dodecanethiol ethanol solution.

Contact-mode AFM was used to pattern the polarization on the ferroelectric substrate, and piezoresponse force microscopy (PFM) was used to confirm the domain orientation. Measurements were performed on a commercial instrument (Digital Instruments Dimension 3000 NS-III). For PFM the AFM was equipped with a function generator and lock-in amplifier (DS340, SRS 830, Stanford Research Systems) and high voltage power supply (PS310, Stanford Research Systems). To protect the electronic system, the electrical connections between the microscope and the tip were severed. A wire was connected from the function generator to the tip using a custom-built sample holder. Internal microscope circuits limit the voltage to 12 Vdc. The modulation amplitude in the PFM imaging was 6 V. Pt- or gold-coated tips (L $\approx$ 125 μm, resonant frequency ~ 350 kHz) (Micromasch NSCS12 W2C) and conductive diamond coated tips (DDESP, Digital Instruments) were used for these measurements.

Nanofabrication involved several steps. The domains in the PZT substrates were patterned by applying +/-10V between an AFM tip and the sample. PFM was used to confirm the orientations of the domains. Samples were submerged in the aqueous metal salt solution and exposed to optical radiation under the range of conditions summarized in Tables I-III. Metal nanoparticles on PZT thin films were characterized by optical microscopy, electron microscopy or AFM after exposure. Samples were then submerged into a solution of functional organic molecules. In order to determine where the organic molecules bonded, x-ray photoemission spectroscopy (XPS) spectra were obtained using a VG Microtech XR3E2 system with a monochromatic Mg Kα X-ray source(1253.6 eV). The energy resolution is about 1 eV.

RESULTS AND DISCUSSION

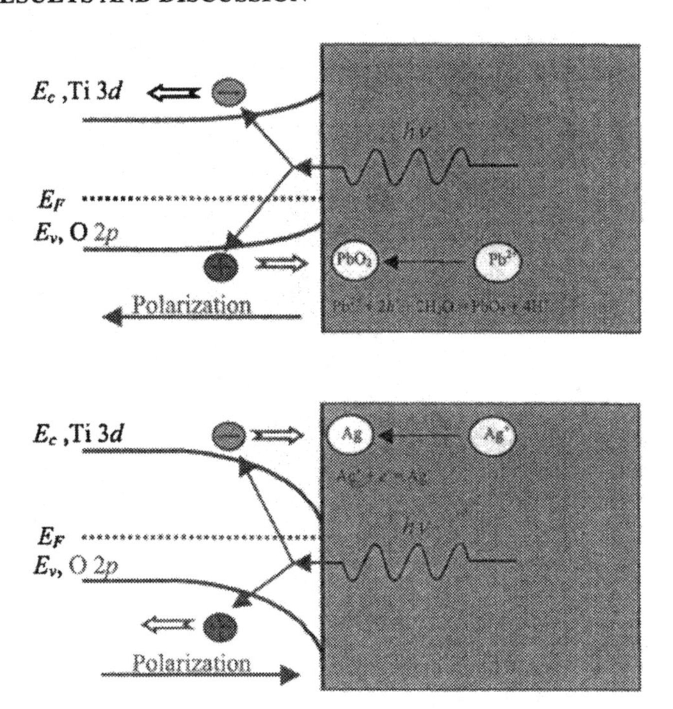

Figure 1. Schematic diagram of near-surface energy levels of atomic orbitals in ferroelectric perovskites showing the influence of polarization at the surface. The relation between band bending and photochemical activity in the c^- (top) and c^+ (bottom) domain regions is illustrated for an oxidation reaction (top) and a reduction reaction (bottom). The response of photo generated electrons and holes to the local fields is indicated by arrows.

Selective deposition of metal nanoparticles on ferroelectric substrates was accomplished for Ag, Au, Pt, and Rh. This paper focuses on those structures

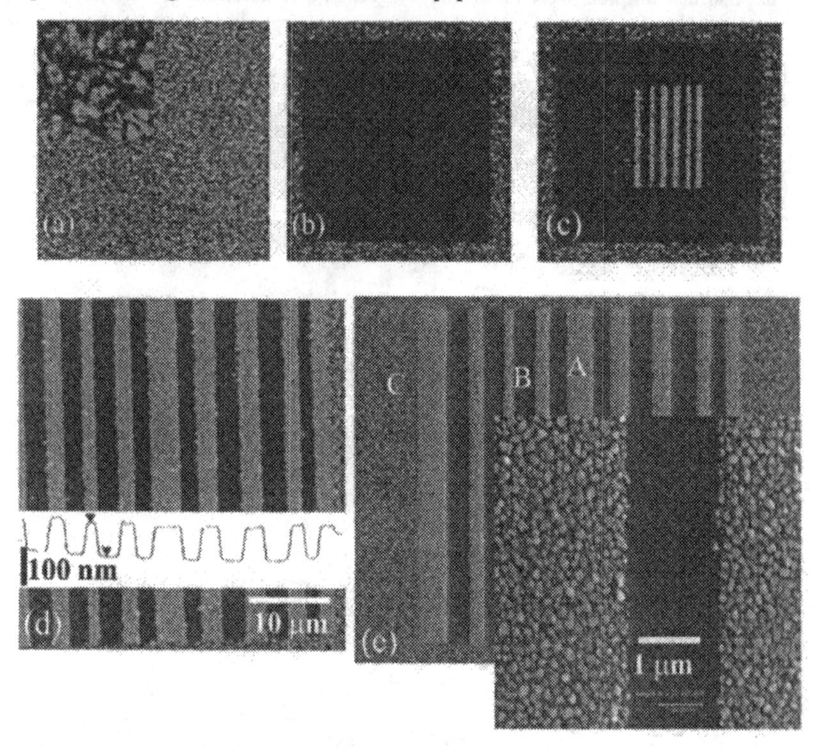

Figure 2. The orientation of atomic polarization in the domains is indicated by the contrast in the Piezoresponse image (a, b, c) of a PZT surface. The bright contrast indicates a positive domain while the dark indicates a negative domain. The surface topography (d, e) after silver deposition shows the location of metal nanoparticles.

based on Ag. The reaction mechanism is based on the relationship between polarization orientation and energies of atomic orbitals at the surface. Describing the electronic structure in terms of a semiconductor, the bottom of the conduction band is predominantly Ti 3*d* orbitals, while the top of the valence band has predominantly O 2*p* character. This relationship is illustrated in Figure 1. Polarization termination at the surface of ferroelectric crystal results in surface charge, the sign of which depends on the orientation of the polarization vector. This charge raises or lowers the energies of the orbitals in the vicinity of the surface, resulting in near-surface band bending and space charge regions. In regions with negative polarization (c^- domains), the effective surface charge becomes more negative and, therefore, upward band bending occurs. In regions with positive polarization (c^+ domains), surface charge is positive, with associated downward band bending. Irradiation with super band-gap radiation results in the formation of electron-hole pairs that are separated by the electric field in the space charge layer. The consequence is that in the absence of an applied electric field, an electron can be donated to a surface reaction only over positive domains, while holes could be donated over negative domains. An electron exchange reaction is then restricted to a domain of a specific orientation.

Figure 2a shows the domain structure at the surface of a PZT thin film. Dark contrast indicates negative polarization; bright contrast indicates positive polarization. Intermediate orientations are also present on unpoled surfaces. A + 10V AFM tip voltage was used to pattern an unreactive square with negative polarization, Figure 2b. Within this, lines of various widths were subsequently patterned by using a −10V biased AFM tip to reorient dipoles to positive orientation immediately below the tip (Figure 2c). The patterned substrates were exposed to 1 x 10^{-3} M Ag nitrate solution under UV light for 15 minutes. Figures 2d and e show the resulting features that exhibit good line definition and demonstrate that these reaction conditions produce a multi layered structure. Above the positive domains (A) silver nanoparticles are close packed with an average size of around 100nm. No particles were observed between the lines (B), i.e. on the negatively charged regions. Outside of the patterned area (C) particles were distributed randomly, consistent with the irregular dipole orientations in unmodified PZT.

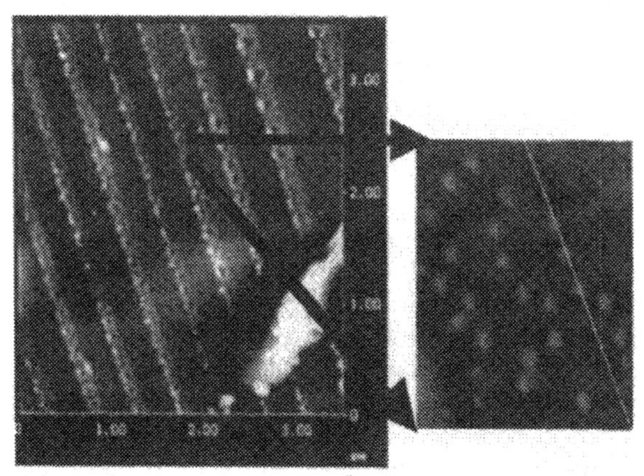

Figure 3 Ag nanoparticles deposited on a $BaTiO_3$ (100) single crystal.

Reaction conditions can be altered to produce less densely packed patterns of smaller particles. This is demonstrated in Figure 3, which shows 3 nm particles along a 100 nm line. The minimum feature size achievable on thin films depends on the grain size of the film. At the small grain sizes found in these films the grains tend to be of a single domain. It is not possible to switch regions smaller than the grains size; therefore the grain size defines the resolution limit. Domain patterning down to 5 nm has been demonstrated on epitaxial films so the fabrication process described here will extend to feature sizes limited by particle size.[25]

Table I Wavelength dependence of photo deposition*

Wavelength (nm)	Energy (ev)	Particle size (nm)	Particle density $(/(\mu m)^2)$	Light intensity (mA)
240	5.17	10	160	4.01
260	4.78	6	120	2.80
280	4.43	12	200	1.55
300	4.14	16	200	1.27
320	3.88	8	24	1.19
340	3.65	5	16	1.09
360	3.45	1.8		1.02
380	3.27	1.6		0.94
PZT		1.5		

*: 1×10-4M silver nitrate aqueous solution, 30minutes irradiation with 300W power.

Table II Concentration dependence of Ag photo deposition*

Concentration (mol/dm^3)	Particle size (nm)	Particle density $(/(\mu m)^2)$
1×10^{-2}	19	115+
1×10^{-3}	18	110+
1×10^{-4}	16	108+
1×10^{-5}	13	62+
1×10^{-6}	8	85

*: 30minutes irradiation with 300W at 300nm wavelength.

Wavelength Dependence of Silver Photoreduction on PZT

In order to quantify the reaction products, the silver particle size and density were determined from topographic AFM images. Because the convolution of the AFM tip significantly affects the lateral measurement of particles, the heights of the particles were used to represent the size by assuming that the shape of most particles within the experiment conditions was nearly spherical. There is significant uncertainty in both the size and density values when the particles are very small due to the background surface roughness. Nevertheless, several reproducible trends are observed. Figure 4(a) shows the wavelength dependence of deposition on 30 minutes exposure in an aqueous solution of 1×10^{-4}M silver

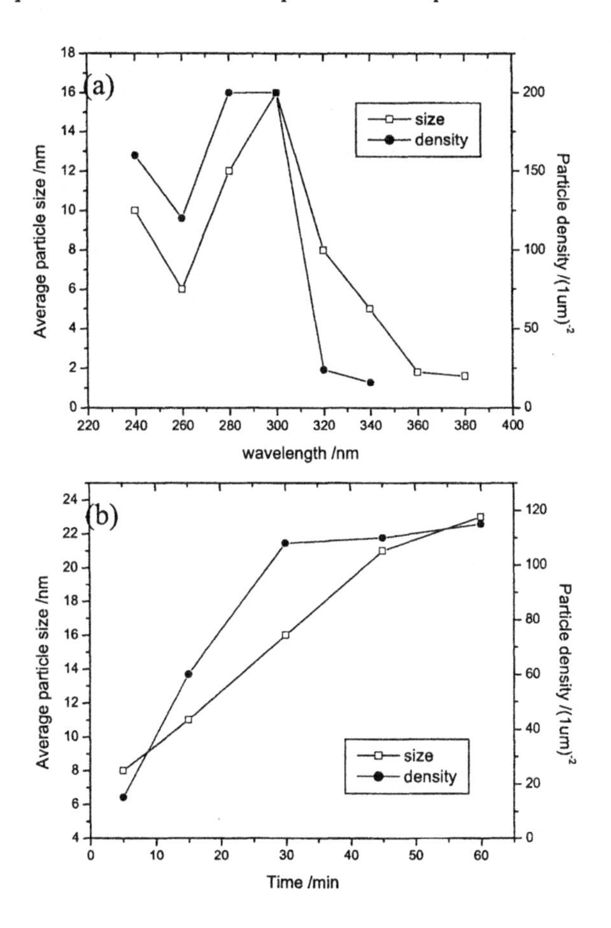

Figure 4. (a) wavelength dependence and (b) time dependence of silver deposition in aqueous solution of 1×10^{-4}M silver nitrate.

nitrate. Optical microscopy detected visible patterns from reactions between 240nm and 340nm but not in the 360nm to 380nm range. AFM measurements showed the average feature size of 1.8nm at 360nm wavelength and 1.6nm at 380nm wavelength, which is very close to that of original PZT thin film, 1.5nm. At 320nm and 340 nm wavelengths, larger particles (8nm and 5nm) and particle density ($24/(\mu m)^2$ and $16/(\mu m)^2$) were obtained. Compared with results obtained from 240nm to 300nm, these are small values. There appears to be no correlation between amount of reaction product and light intensity (See table I). The maximum in size and particle density, and therefore reaction product occurs at 4.1 ev, i.e. 0.5 eV above the substrate band gap, 3.6 eV for PZT. The uncertainty in distinguishing metal clusters from surface roughness for the smallest particles precludes conclusions regarding nucleation and growth rates.

Kinetics of Silver Photoreduction on PZT

Figure 4(b) compares the evolution of particle size and density on exposure to 300nm wavelength light in an aqueous solution of 1×10^{-4}M silver nitrate. As the reaction time increased from 5 to 60 minutes, the particle size increased monotonically a factor of 3 and the density increased a factor of 6. It appears that the particle density saturates at about 30 minutes, while the particle size continues to grow. The uncertainty in the measurement and the necessity of relying on average particle size rather than particle size distribution preclude calculating the nucleation and growth rates. It does appear though that nucleation decreases substantially at very low surface coverages, as indicated in figure 4 (b).

Table II illustrates the concentration dependence after exposure to 300nm wavelength light in an aqueous solution of silver for 30 minutes. Increasing the concentration by 5 orders of magnitude results in increasing particle size and density by a factor of 2. The particle density reported in Table II is an under representation when accompanied by "+", which indicates particle spacing made measurement difficult. It is interesting to note that since the PZT film grain size is 50-100nm, at the final density in these experiments each grain was occupied by one big 10~20 nm silver particle on average.

Photoreduction Mechanism

It is clear that several processes participate in the photoreduction reaction, one at the initial stages and one at later stages. A silver ion in an aqueous environment will be hydrated or hydroxylated. The complex resulting from hydroxylation would behave as a negative charge. A hydrated ion might expose a positive charge. These complexes would separate in a spatially inhomogeneous diffuse double layer that compensates the ferroelectric domains. As described in Figure 1, reduction will occur only over the positive domains. An electron is transferred from the Ti band to an adsorbed ion, probably through a Ti-O-Ag+ chain producing silver atoms at the surface. Nucleation of clusters can occur via diffusion of metal atoms on the surface or reaction at adjacent sites. Sizes of the particles at later stages of the reaction and the density and growth kinetics in figure 4 (b) indicate that reduction also occurs on the metal. In this case electrons from the substrate are injected into the particle to the silver ion at the surface. We have also found reduction can occur though a thin film of amorphous carbon, in which the electrons must also traverse an intermediate metallic structure. Details of these processes remain to be determined.

The 500 meV above the energy of the band gap required for the reaction may be a consequence of the interface transport. We also found almost same energy above the bandgap was needed for gold photoreduction (reagent was $HauCl_4$). As one Au^{3+} needs three electrons to be reduced to Au^0, the gold ion unit ($[AuCl_4]^-$) may play an important role.

Complex Structure Fabrication

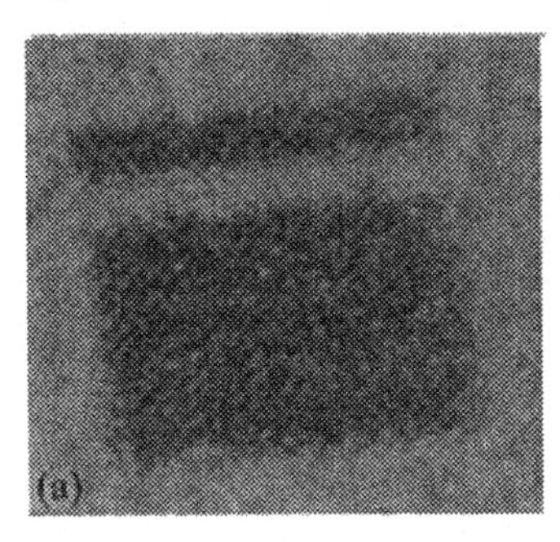

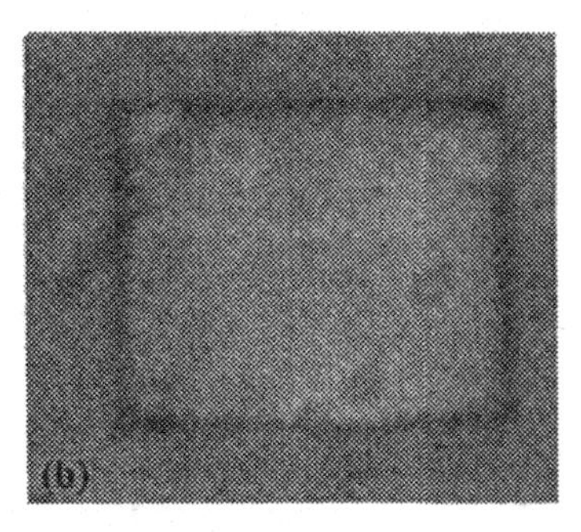

Figure 5. Optical micrographs of deposition of multiple metal particles. The color contrast is due to various thickness of the deposited metals. Sequential deposition of Ag and Au from nitrate(a), simultaneous deposition of Ag and Au from nitrate (b). Presence of metal was determined by energy dispersive spectroscopy in an SEM.

The process was extended to deposit multicomponent nanoparticles. Nanoparticles of multiple metals were deposited by two processes. In the first, sequential exposure in the presence of two separate metal salt solutions, for example $AgNO_3$ then $HAuCl_4$, resulted in layers of Au nanoparticles on top of Ag nanoparticles. The second process involved a single exposure to a solution of multiple metal ions, for example mixed $AgNO_3$ and $HauCl_4$. The bi metallic patterns in figure 5 were observed by SEM and the presence of both metals was confirmed by EDS.

Ferroelectric lithography and photodeposition result in composite structures of insulating or semiconducting substrates and metal nanoparticles in predefined locations. To make more complex structures, these patterns can be further reacted with organic or biological molecules. Taking advantage of reactions used in organic self-assembly, such as thiol-gold or carboxylic acid-oxide couplings, molecules can be attached to the nanoparticles. This was illustrated in a model system with PZT/Ag/dodecanethiol. A substrate patterned with Ag nanoparticles was exposed to 1×10^{-4} M dodecanthiol of ethanol solution for 24 hours. X ray photoemission spectroscopy was used to track the sulfur. It was found that sulfur, and therefore the organic molecule is attached only to the metal particles. This is determined from a comparison of the PZT/Ag/thiol with control samples (clean PZT exposed to thio, PZT exposed to metal salt solution then thiol but no light). The $Ag_{3d}5/2$ peak and $Ag_{3d}3/2$ peak at 168 eV and 174 eV respectively were resolved. The S_{2p} spectra at 161.5 ~ 163.5 eV are identified as sulfur species bound to Ag particles. This is distinguished from the S_{2p} peak at 168 eV which is identified as sulfur bound to O on the PZT substrate.

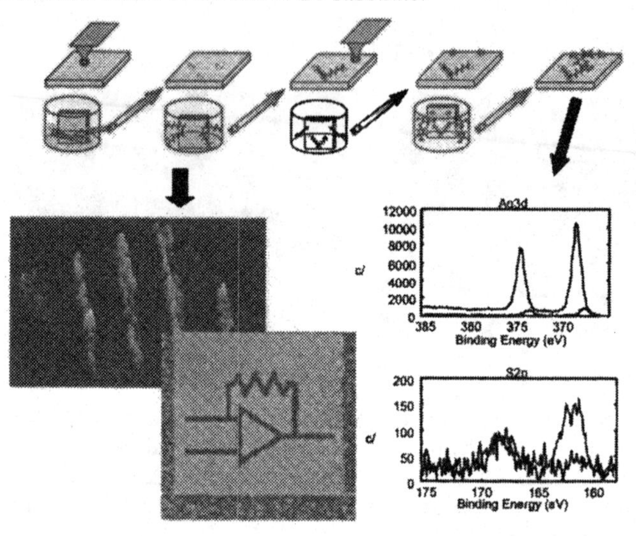

Figure 6. A schematic diagram of the ferroelectric nanolithography of complex structures. The sequence involves polarization patterning, photoreduction of metal particles, covalent attachment of functional organic molecules in a series that can be repeated to add nanostructural components. Patterning is confirmed in PFM images, AFM shows nano particles positioned in a line. The attachment of organic molecules on the metal particles is confirmed by XPS which detects the sulfur atom. Control samples with no metal particles contain only small amounts of S and ionic Ag adsorbed from solution.

In order to fabricate more complex structures these processes can be repeated sequentially. Subsequent iteration can utilize different metal nanoparticles, different molecules, etc. (Figure 6). This method can also be combined with Si technology to be integrated. It is possible to fabricate a patterned electrode with conventional Si fabrication technology or with microcontact printing technique. This approach is parallel in feature production rather than serial. Functional nano constituents might be nanotubes, nanowires, polypeptides and other organic molecules.

CONCLUSION

These results demonstrate a novel approach for the assembly of complex nanostructures. Controlled polarization switching with subsequent metal photodeposition allows creation of metal meso- and nanoscale structures. Polarization and deposition steps can be repeated thus allowing fabrication of nanostructures comprised of several deposited materials on a substrate. It is important to note that this mechanism of directed assembly differs fundamentally from those that utilize local electrostatic attraction to assemble nanostructures onto templates of patterned charge.[26] In the latter case, local charge can be used to locally deposit charged particles from colloidal solution. In these cases, the positions of the charges are not pinned; therefore, the pattern is susceptible to diffusion. On a ferroelectric substrate, local surface charge is due to atomic polarization and therefore is stable. More importantly, since the reaction mechanism involves controlling the surface electronic structure, the reaction product is not limited by the amount of local charge. Through adjusting the photoreduction conditions, composition and morphology of nanoparticles can be varied to satisfy design criteria. When combined with chemical assembly nanostructures consisting of oxide substrates, metal nanoparticles and organic/biological molecules can be fabricated. The procedure can be repeated to fabricate structures in which multiple types of metal particles and several electronically or optically active molecules can be assembled in predetermined configurations, paving the way to molecular electronic, optoelectronic and chemical sensor devices with multiple functionalities.

ACKNOWLEDGEMENTS

We acknowledge the support from NSF grant DMR 96-32596 for facilities use, the Center for Science and Engineering of Nanoscale Systems (SENS) at the University of Pennsylvania, and the Department of Education. We have benefited from extensive discussions with John Vohs, and S.Dunn.

REFERENCE

[1]D. V. Averin, K. K. Likharev, "Coulomb Blockade of Single-Electron Tunneling, and Coherent Oscillations in Small Tunnel Junctions", J. Low Temp. Phys., 62, 345 (1986).

[2]D. C. Ralph, C. T. Black, M. Tinkham, "Gate-voltage studies of discrete electronic states in aluminum nanoparticles", Phys. Rev. Lett., 78, 4087-4090 (1997).

[3]M. Fleischmann, P. J. Hendra, A. J. McQuillan, "Raman Spectra of Pyridine Adsorbed at a Silver Electrode", Chem. Phys. Lett., 26, 163 (1974).

[4]M. F. Crommie, C. P. Lutz, D. M. Eigler, "Confinement of Electrons to Quantum Corrals on a Metal Surface", Science, 262, 218-220 (1993).

[5]M. Asakawa, M. Higuchi, G. Mattersteig, T. Nakamura, A. R. Pease, F. M. Raymo, T. Shimizu, J. F. Stoddart, "Current/voltage characteristics of monolayers of redox-switchable [2]catenanes on gold", Adv. Mater., 12, 1099-1102 (2000).

[6]C. P. Collier, G. Matterstei, E. W. Wong, Y. Luo, K. Beverly, J. Sampaio, F. M. Raymo, J. F. Stoddart, J. R. Heath, "A [2]catenane-based solid state electronically reconfigurable switch", Science, 289, 1172-1175 (2000).

[7]D. L. Simone, T. M. Swager, "A Conducting Poly(cyclophane) and Its Poly([2]-catenane)", J. Am. Chem. Soc., 122, 9300-9301 (2000).
[8]V. Balzani, A. Credi, S. J. Langford, F. M. Raymo, J. F. Stoddart, M. Venturi, "Constructing Molecular Machinery: A Chemically-Switchable [2]Catenane", J. Am. Chem. Soc., 122, 3542-3543 (2000).
[9]C. P. Collier, E. W. Wong, M. Belohradsky, F. M. Raymo, J. F. Stoddart, P. J. Kuekes, R. S. Williams, J. R. Heath, "Electronically configurable molecular-based logic gates", Science, 285, 391-394 (1999).
[10]J. M. Tour, "Molecular Electronics. Synthesis and Testing of Components", Acc. Chem. Res., 33, 791-804 (2000).
[11]J. Chen, W. Wang, M. A. Reed, A. M. Rawlett, D. W. Price, J. M. Tour, "Room-temperature negative differential resistance in nanoscale molecular junctions", Appl. Phys. Lett., 77, 1224-1226 (2000).
[12]J. Chen, M. A. Reed, A. M. Rawlett, J. M. Tour, "Large on-off ratios and negative differential resistance in a molecular electronic device", Science, 286, 1550-1552 (1999).
[13]I. D. Norris, M. M. Shaker, F. K. Ko, A. G. MacDiarmid, "Electrostatic fabrication of ultrafine conducting fibers: polyaniline/polyethylene oxide blends", Synth. Met., 114, 109-114 (2000).
[14]A. G. MacDiarmid, A. J. Epstein, "The polyanilines: a novel class of conducting polymers", Mater. Res. Soc. Symp. Proc., 173, 283 (1990).
[15]G. Y. Tseng, J. C. Ellenbogen, "Toward Nanocomputers", Science, 294, 1293-1294 (2001) and references therein.
[16]P. C. Hidber, W. Helbig, E. Kim, G. M. Whitesides, "Microcontact printing of palladium colloids: Micron-scale patterning by electroless deposition of copper", Langmuir, 12, 1375-1380 (1996).
[17]H. X. He, H. Zhang, Q. G. Li, T. Zhu, S. F. Li, Z. F. Liu, "Fabrication of designed architectures of Au nanoparticles on solid substrate with printed self-assembled monolayers as templates", Langmuir, 16, 3846-3851 (2000).
[18]T. Junno, S. B. Carlsson, H. Xu, L. Montelius, L. Samuelson, "Fabrication of quantum devices by angstrom-level manipulation of nanoparticles with an atomic force microscope", Appl. Phys. Lett., 72, 548-550 (1998).
[19]R. Resch, A. Bugacov, C. Baur, B. E. Koel, A. Madhukar, A. A. G. Requicha, P. Will, "Manipulation of nanoparticles using dynamic force microscopy: simulation and experiments", Appl. Phys. A, 67, 265 (1998).
[20]H. Sugimura, N. Nakagiri, "Nanoscopic surface architecture based on scanning probe electrochemistry and Molecular self-assembly", J. Am. Chem. Soc., 119, 9226-9229 (1997).
[21]Q. Li, J. W. Zheng, Z. F. Liu, "Site-selective assemblies of gold nanoparticles on an AFM tip-defined silicon template", Langmuir, 19, 166-171 (2003).
[22]K. R. Brown, M. J. Natan, "Hydroxylamine seeding of colloidal Au nanoparticles in solution and on surfaces", Langmuir, 14, 726-728 (1998).
[23]F. Jona and G. Shirane, in Ferroelectric Crystals, Dover Publications, New York 1993.

[24]J.L. Giocondi, G.S. Rohrer, “Spatially Selective Photochemical Reduction of Silver on the Surface of Ferroelectric Barium Titanate”, Chem. Mater., 13, 241-242 (2001).
[25]J. Junquera, P. Ghosez, “Critical thickness for ferroelectricity in perovskite ultrathin films”, Nature, 422, 506-509 (2003).
[26]P. Mesquida, A. Stemmer, “Attaching Silica Nanoparticles from Suspension onto Surface Charge Patterns Generated by a Conductive Atomic Force Microscope Tip”, Adv. Mat., 13, 1395-1398 (2001).

INFLUENCES OF ADDITIVES ON THE FORMATION OF THIN ZnO FILMS ON SELF-ASSEMBLED MONOLAYERS

Rudolf C. Hoffmann, Joachim Bill and Fritz Aldinger

Max-Planck-Institut für Metallforschung and Institut für Nichtmetallische Anorganische Materialien, Universität Stuttgart, Pulvermetallurgisches Laboratorium, Heisenbergstr. 3, 70569 Stuttgart, Germany

ABSTRACT

The deposition of homogeneous and adherent ZnO films on Si wafers by thermohydrolysis of zinc salts in aqueous solution is reported. Using a graft copolymer (polymethacrylic acid partially grafted with polyethyleneoxide side chains) homogeneous films of nanosized particles on Si wafers with self-assembled monolayers (SAMs) were obtained. The addition of the polyelectrolyte is necessary to suppress the formation of undesired, larger zincite crystals, which form inevitably otherwise.

The deposition of films was investigated on surfaces, which were functionalized with mercapto- or sulfo-terminated SAMs, as well as on unfunctionalized substrates. The morphology of the films was characterized by light microscopy and atomic force microscopy. Only incomplete coverage of the substrate surface was achieved on SAMs with sulfo groups or on unfunctionalized surfaces, whereas complete coverage of the substrates was obtained after modification with mercapto functions.

INTRODUCTION

The synthesis of materials and devices from aqueous solutions at low temperature is an emerging field of research.[1] For the deposition of ceramic thin films a variety of reaction procedures were developed in the last years, such as Chemical Bath Deposition (CBD), Liquid Phase Deposition (LPD) or Electroless Deposition (ELD).[2,3] A crucial factor for the deposition process regardless of the applied method is the surface chemistry of the substrate. Partially inspired by the strong degree of control in biomineralization, the idea of introducing an intermediate organic layer between the substrate and the ceramic film was developed. For this purpose so-called self-assembled monolayers (SAMs) are now widely applied. For ceramic surfaces trichloro- or trialkoxyorganosilanes were successfully used to built up SAMs.[2,3]

The deposition of ZnO however encounters several difficulties. The growth of ZnO thin films by CBD is well investigated. Hereby commonly aqueous solutions of zinc salts are hydrolyzed at temperatures between 50 and 100°C. The films obtained on silica glass or tin doped indium oxide (ITO) is deposited in the form of oriented, rather large micron-sized zincite crystals.[4,5] Although such films show interesting optical properties, the large crystal size limits the resolution for micropatterning severely. Furthermore unfavorable reaction conditions lead to agglomerations of crystals with low adherence and a high surface roughness.[4,5]

Recently, the deposition of ZnO thin films by ELD was reported. Films were obtained from solutions of zinc nitrate and dimethylaminoborane. Hereby the nitrate ion is reduced by the borane and a subsequent reaction between the resulting hydroxide ions and zinc ions lead to the deposition. ZnO thin film were generated on photopatterned SAMs from phenyltrichlorosilane at 55°C. Although the course of the reaction is slightly different to CBD, the resulting films also consist of micron-sized crystals.[6]

Zinc oxide, which is precipitated in bulk or grown on films has an explicit tendency to form long zincite needles. Andrès-Vergès *et al.* investigated the morphology of ZnO particles obtained from bulk precipitation at different times after the onset of precipitation. In the first 3–7 min spherical particles were formed, which would be the desired building material for thin films, but after 10 min the product consisted only of micrometer-sized hexagonal needles.[7]

One way to overcome this problem could be the addition of a dispersant to the reaction solution in order to stabilize the nanoparticles, that are formed at an early stage of the thermohydrolysis. Previously, Wegner and co-workers successfully used polyethylenoxide-block-polymethacrylic acid (PEO-block-PMAA) copolymers to control the morphology of ZnO bulk precipitation. The PMAA unit interacts with the surface of the ZnO and functions as an anchor group. The PEO moiety does not bind to the surface and acts as a steric shield. Block and also graft copolymers with such functionalities should be able to stabilize ZnO nanoparticles and prevent them from rapid growth.[8]

Recently, we successfully used PEO-graft-PMAA copolymers to obtain ZnO thin films without the deposition of micron-sized crystals. The reaction system, which was applied comprises a number of parameters, which have to be adjusted in order to ensure the requested film morphology. So far our investigations concentrated on the composition of the reaction solution. We report here the influence of different SAM surface groups on the film morphology and the formation process.[9,10]

EXPERIMENTAL PROCEDURES

SAM deposition: Surface functionalization of silicon p-type {001} wafers with γ-mercaptopropyltrimethoxysilane was carried out according to published procedures. To convert the mercapto (-SH) into a sulfo ($-SO_3H$) functionality, the wafers were immersed in oxone™ (potassium hydrogenmonopersulfate) for a minimum of 4 h at room temperature.[9,11]

Silicon wafers, which were not coated with SAMs, were surface oxidized with a mixture of sulfuric acid and hydrogen peroxide (v:v, 70:30) for comparison with modified samples, which had to be hydroxylated before the SAM deposition.[9,11]

Film deposition and characterization: The graft-copolymer employed was the ammonium salt of poly[methacrylic acid-co-(methacrylic acid-graft-polyethylenoxide)] $P(MAA_{0.66}\text{-co-}(MAA\text{-}EO_{20})_{0.33})_{70}$ (M_n = 17.600 $g\cdot mol^{-1}$).

The deposition solution contained 100 mM [Zn] (80 mM $ZnCl_2 \cdot 6\ H_2O$ and 20 mM $Zn(CH_3COO)_2 \cdot 6\ H_2O$), 2500 ppm copolymer and 2.5 mM hexamethylenetetramine (HMTA). To ensure optically clear starting solutions the components were mixed according to the sequence published earlier.[9]

The substrates were immersed in 10 ml aliquots of the deposition solution, covered and then placed in an oil bath at 80°C. The solution was exchanged after 2 h. Total deposition times were 4–8 h. The pH of the solution started at 6.4 and was 6.6–6.7 afterwards. No visible bulk precipitation occurred during such depositions. Finally, the samples were washed with distilled water, ultrasonicated and dried in a stream of dry argon.

Annealed samples were obtained by heating at 723 K for 30 minutes in air (heating rate 10 K/h, cooling rate 40 K/h).

Films were investigated by light microscopy (LM) and atomic force microscopy (AFM) using a Digital Instruments Nanoscope III in tapping mode with silicon cantilevers.

Zeta potential Measurements: A Zetasizer Malvern 3000 HS_A was used to measure the electrophoretic mobility. All measurements were carried in an aqueous solution of 1 mM aqueous KNO_3 at 25°C. The diluted suspension of ZnO (Alfa, d_{50} 0.780 µm, S_{BET} 5.788 $m^2\cdot g^{-1}$) was sonicated for deagglomeration. The first measurement was carried out at the intrinsic pH value of the ZnO suspension. The pH was then changed by addition of a 250 mM KOH solution or the dropwise addition of a concentrated aqueous solution of the PMAA-graft-PEO copolymer, respectively. To improve the reliability, a series of three separate measurements was collected for each system.

RESULTS AND DISCUSSIONS

Although the film deposition was carried out with the same composition of the reaction solution in all cases, the morphologies of the films, which formed, were significantly different. Complete macroscopic coverage of the surface was achieved only with mercapto-functionalized wafers. These films remained adherent also after annealing.[9] Only partial coverage (and sometimes even no coverage at all) was observed on unfunctionalized or sulfo-functionalized silicon wafers (Fig. 1). AFM images of the covered areas on sulfo-functionalized silicon wafers revealed large holes, which remain uncovered even after longer deposition times. Judging from AFM measurements films grown on unfunctionalized or sulfo-functionalized wafers had a thickness of about 60±10 nm after 4 hours and about 120±10 nm after 8 hours, suggesting an average growth rate of about 15 nm/hour. The depth of the holes equals the thickness of the films (Fig. 2).

Figure 1: LM images of samples deposited during 4 hours on differently functionalized surfaces. (a) without SAM, (b) with sulfo SAM, (c) with mercapto SAM and (d) deposited on mercapto SAM and calcined after film deposition.

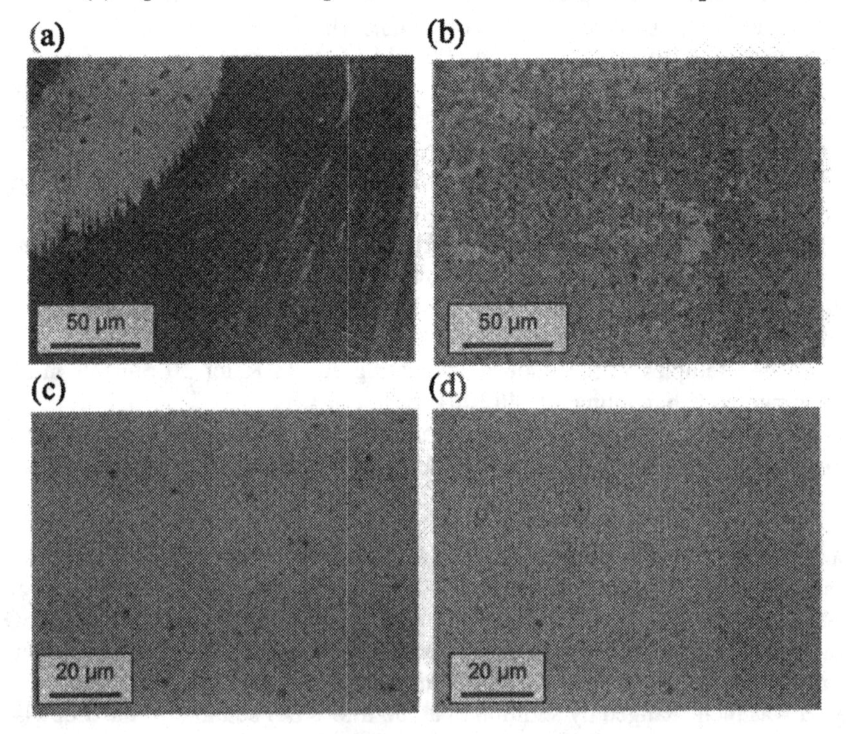

Surprisingly though the mercapto-functionalized silicon wafers were macroscopically completely covered, AFM images revealed a canyon-like morphology with a high surface roughness (Fig 3.). As discussed in a recent publication this surface morphology occurs at higher grow rates and can be avoided by controlling the extent of the pH-shift, i.e. the amount of HMTA used in the reaction.[10]

Figure 2: AFM image of as-deposited film grown on silicon wafer without SAM modification. (a) after 4 hours (Z = 161.5 nm) and (b) after 8 hours (Z = 168.6 nm).

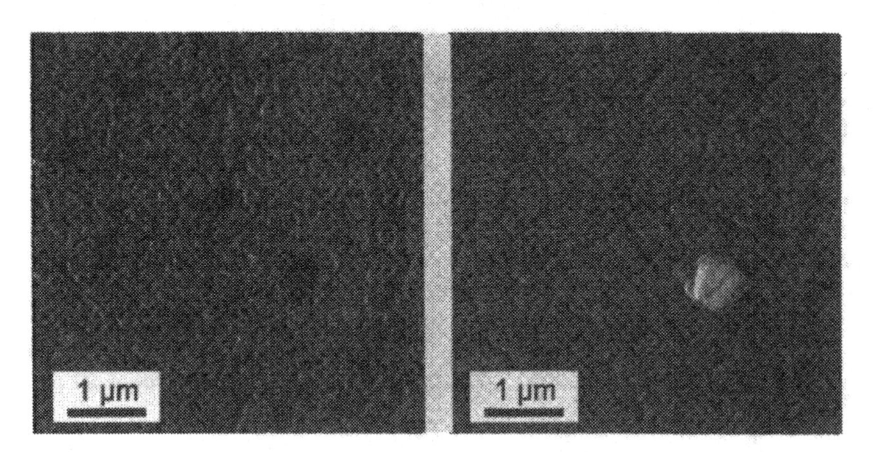

Figure 3: AFM image of as-deposited film grown on silicon wafer with SH-SAM. (a) after 6 hours (Z = 136.9 nm) and (b) after 8 hours (Z = 163.4 nm).

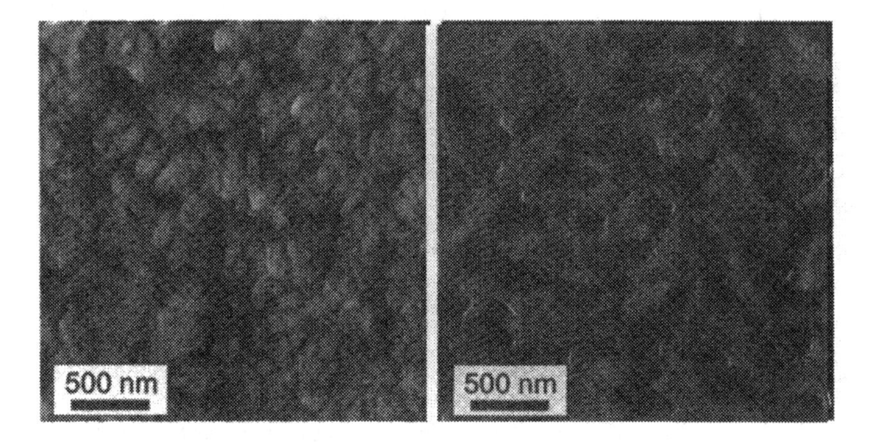

According to AFM measurements the thickness of these film (Fig. 3) on the mercapto-functionalized surface is comparable to that of unfunctionalized (Fig. 2) or sulfo-functionalized wafers. It must be emphasized though, that as mentioned above in two latter cases no complete coverage of the silicon wafer was obtained.

The films are amorphous to X-Ray and electron diffraction. Earlier investigations by X-Ray photoelectron spectroscopy suggested, that they consist of ZnO particles, which are capped with copolymer molecules. To clarify the surface charge of the particles under the reaction conditions the zeta potential of

ZnO suspensions with and without the copolymer was determined (Fig. 4). This model system was used, since measurements of the reaction solution itself, proved to be difficult so far.

The curve of ZnO without the addition of copolymer is in accordance with earlier publications, revealing an isoelectric point (IEP) of about 9.0.[12] The copolymer shifts the curve to lower pH values (Fig. 4). Similar observations were reported for slurries of Si_3N_4 or SiC in ethanol in the presence of graft-copolymers, which were investigated for their use as dispersants.[13] From these measurements it can be assumed, that the ZnO/copolymer particles are slightly negatively charged in the film deposition experiments.

Figure 4: Zeta potential of ZnO in water as a function of pH. Operational pH was controlled by addition of KOH (rhombs) or PMAA-graft-PEO copolymer (squares). The pH regime of the film deposition is indicated by the hatched area.

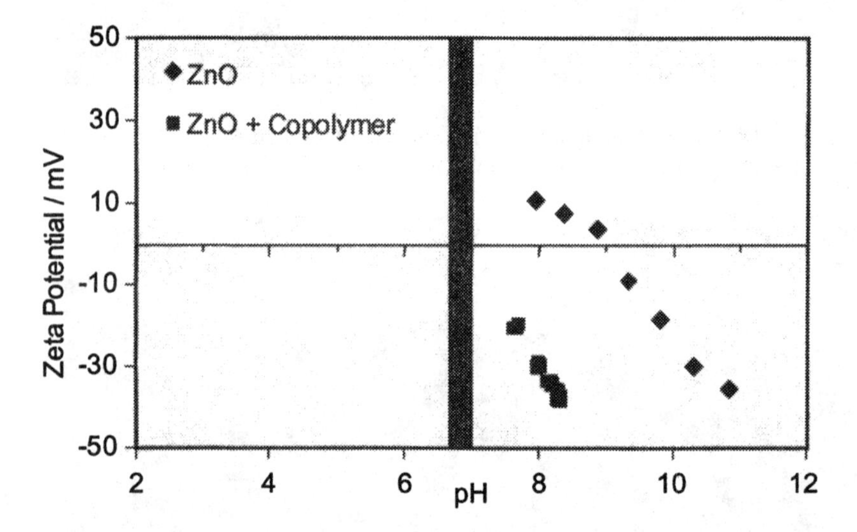

CONCLUSIONS

The groups of *De Guire* and *Sukenik* investigated the deposition of titania films by LPD on a variety of substrates and at various pH values. A working model was established for the attachment of colloidal particles formed in solution by the CBD or LPD process onto SAM-modified surfaces. As a rule of thumb the surface charge of the particles and the SAMs, which can be both estimated from zeta potential measurements, must lead to attractive interactions.[11] The suggested mechanism is supported by the DLVO theory (Derjaguin, Landau, Verwey and Overbeck), which quantifies electrostatic and van der Waals interactions as well contributions from other forces between the precipitate and the SAM.[14]

In agreement with our earlier investigations we propose, that ZnO nanoparticles are formed in the course of the thermohydrolytic reaction, which are capped by PMAA-graft-PEO molecules. At the pH value of the deposition reaction such particles are slightly negatively charged and thus would preferably deposit on positively charged surfaces. However van der Waals interactions between the ZnO/copolymer particles and the surface can not be neglected. Further bridging and flocculation effects caused by the copolymer molecules might play a role in the deposition process.

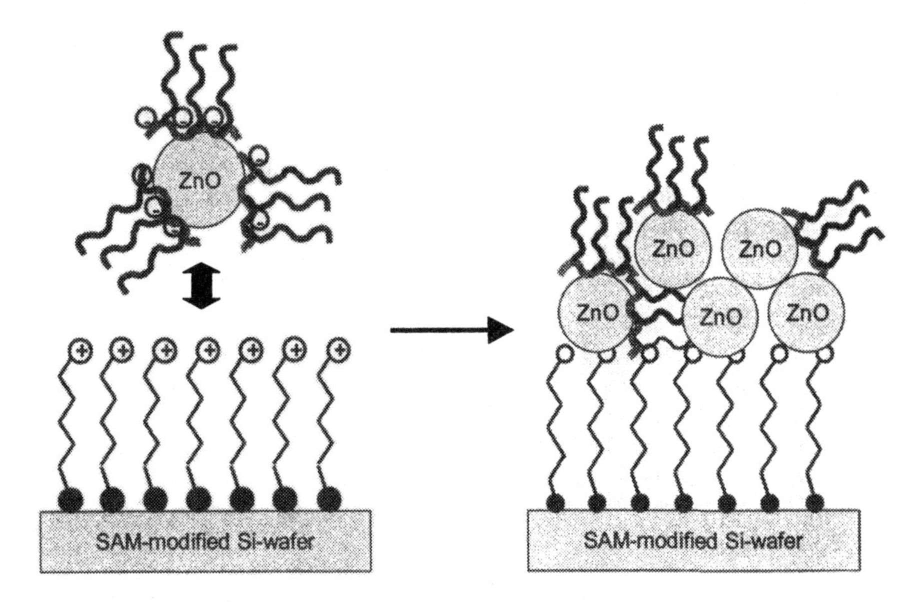

Figure 5: Schematic presentation of the film formation from ZnO particles stabilized with graft-copolymers.

ACKNOWLEDGEMENT

This project was supported by the DFG through grants AL 384/29-1 and AL 384/22-3 and the BMBF through grant 03C0294C/8. The assistance of Ms Stefanie Wildhack in obtaining the zeta potential data is highly appreciated.

REFERENCES

[1]M. Yoshimura, W. Suchanek and K.S. Han, "Recent developments in soft, solution processing: One step fabrication of functional double oxide films by hydrothermal-electrochemical methods", *Journal of Materials Chemistry*, **9** [1] 77-82 (1999).

[2]J. Bill, R.C. Hoffmann, T.M. Fuchs and F. Aldinger, "Deposition of ceramic materials form aqueous solutions induced by organic templates", *Zeiftschrift für Metallkunde*, **93** [5] 479-489 (2002).

[3]T.P. Niesen and M.R. De Guire, "Review: Deposition of ceramic thin films grown at low temperatures from aqueous solution", *Journal of Electroceramics*, **6** [3] 169-207 (2001).
[4]D.S. Boyle, K. Govender and P. O'Brien, "Novel low temperature solution deposition of perpendicularly orientated rods of ZnO: substrate effects and evidence of the importance of counter-ions in the control of crystal growth", *Chemical Communications*, **15** 80-81 (2002)
[5]K. Govender, D.S. Boyle, P. O'Brien, D. Binks, D. West and D. Coleman, "Room-temperature lasing observed from ZnO nanocolumns grown by aqueous solution deposition", *Advanced Materials*, **14** [17] 1221-1224 (2002).
[6]N. Saito, H. Haneda, T. Sekiguchi, N. Ohashi, N. Ohashi, I. Sakaguchi and K. Koumoto, "Low-temperature fabrication of light-emitting zinc oxide micropatterns using self-assembled monolayers", *Advanced Materials*, **14** [6] 418-421 (2002).
[7]M. Andrés Vergés, A. Mifsud and C.J. Serna, "Formation of rod-like zinc oxide microcrystals in homogeneous solutions", Journal of the Chemical Society Faraday Transactions, **86** [6] 959-963 (1990).
[8]M. Öner, J. Norwig, W.H. Meyer and G. Wegner, "Control of ZnO crystallization by a PEO-b-PMAA diblock copolymer" *Chemistry of Materials*, **10** [2] 460-463 (1998).
[9]R. Hoffmann, T. Fuchs, T.P. Niesen, J. Bill and F. Aldinger,"Influence of PMAA-graft-PEO copolymers on the formation of thin ZnO films from aqueous solutions", *Surface and Interface Analysis*, **34** [1] 708-711 (2002).
[10]R.C. Hoffmann, S. Jia, J.C. Bartolomé, T.Fuchs, J. Bill, P.C.J. Graat and F. Aldinger, "Growth behaviour of thin ZnO films from aqueous solutions in the presence of PMAA-graft-PEO copolymers", *Journal of the European Ceramic Society*, accepted.
[11]H. Pizem, C.N. Sukenik, U. Sampathkumaran, A.K. Mc Ilwain and M.R. De Guire, "Effects of substrate surface functionality on solution-deposited titania films", *Chemistry of Materials*, **14** [6] 2476-2485 (2002).
[12]A. Degen and M. Kosec, "Effect of pH and impurities on the surface charge of zinc oxide in aqueous solution", *Journal of the European Ceramic Society*, **20** [6] 667-673 (2000).
[13]E. Laarz and L. Bergström, "The effect of anionic polyelectrolytes on the properties of aqueous silicon nitride suspensions", *Journal of the European Ceramic Society*, **20** [4] 431-440 (2000).
[14]Y. Tang, Y. Liu, U. Sampathkumaran, M.Z. Hu, R. Wang and M.R De Guire, "Particle growth and particle-surface interactions during low-temperature deposition of ceramic thin films", *Solid State Ionics* **151** [1-4] 69-78 (2002).

Fabrication of Metallic Nanocrystal Arrays for Nanoscale Nonlinear Optics

Anthony B. Hmelo
Dept. of Physics and Astronomy
Vanderbilt University
Nashville, TN 37235

Matthew D. McMahon
Dept. of Physics and Astronomy
Vanderbilt University
Nashville, TN 37235

Rene Lopez
Dept. of Physics and Astronomy
Vanderbilt University
Nashville, TN 37235

Robert H. Magruder, III
Chemistry and Physics Department
Belmont University
Nashville, TN 37212

Robert A. Weller
Department of Electrical
Engineering and Computer Science
Vanderbilt University
Nashville, TN 37235

Richard F. Haglund, Jr.
Dept. of Physics and Astronomy
Vanderbilt University
Nashville, TN 37235

Leonard C. Feldman
Dept. of Physics and Astronomy
Vanderbilt University
Nashville, TN 37235

ABSTRACT

Ordered arrays of metal nanocrystals embedded in or sequestered on dielectric hosts have potential applications as elements of nonlinear or near-field optical circuits, as sensitizers for fluorescence emitters and photodetectors, and as anchor points for arrays of biological molecules. Here we report on the fabrication of two-dimensional ordered metallic nanocrystal arrays for optical investigations. We have employed strategies for synthesizing metal nanocrystal composites that capitalize on the best features of focused ion beam (FIB) machining and pulsed laser deposition (PLD) of Ag on Si/SiO_2 substrates.

INTRODUCTION

Ordered arrays of metal nanoclusters have potential applications as elements of nonlinear or near-field optical circuits, as sensitizers for fluorescence emitters and photodetectors, and as anchor points for arrays of biological molecules. Metal nanocrystals are strongly confined electronic systems with a band structure drastically altered by the small size of the system and the reduced population of conduction-band electrons. Their optical response is extremely sensitive to the size, size distribution and spatial arrangement of individual nanocrystals. This effort is focused on the fabrication of ordered metallic nanocrystal arrays for optical investigations. Our approach is to promote the formation of metal nanocrystals on a spatially arranged array of holes that has been prepared by focused ion beam (FIB) processing.

Periodic arrays of metal nanocrystals embedded in a dielectric have several interesting properties. The property we wish to exploit is the third-order nonlinear optical response of these arrays, which gives rise to such effects as intensity-dependent index of refraction and nonlinear absorption. Ordered arrays using this property could be used in photonic circuits, particularly in interferometry schemes and possibly in all-optical switching. The purpose of the experiments described in this paper is to create ordered arrays of Ag nanocrystals on oxidized silicon for such optical experiments.

EXPERIMENTAL

The substrate is a highly conductive n-type silicon, with a surface oxide layer thickness of 37 nm determined by ellipsometry. The "metallic" n-type silicon single crystal substrates were first etched in HF to remove native surface oxide. The wafers were subsequently re-oxidized in a controlled furnace environment to prepare the final oxide layer.

We use a FIB system to create arrays of sites at which metal nanocrystals are expected to nucleate and grow following the deposition of a metal layer. The principles of material removal via sputter erosion using the FIB are well known [1, 2]. Facilities at the Vanderbilt Nanofabrication Laboratory include the FEI FIB 200, which features a 30kV liquid Ga^+ ion source. In the FIB ion source liquid gallium is drawn to a Taylor cone from which individual ions can be extracted. The ions are accelerated through 30kV potential and focused onto the substrate, where material is sputtered from the surface. The FIB is computer-controlled, and patterns of arbitrary complexity can be written to the substrate with user-defined pixel-addressable pattern files. The smallest line width (FWHM of the FIB) is 8 nm, according to manufacturer specifications; we have demonstrated 10 nm features. The experimental approach is to create ordered damage sites in Si/SiO_2 and other substrates using the FIB, and to subsequently use deposition techniques to form nanoparticles, with the expectation that the particles will nucleate and grow at the selected sites.

The Ag thin film layer was made by pulsed laser deposition (PLD). PLD is a versatile technique for depositing and growing nanoparticles and thin films of metals, semiconductors and insulators [3]. In PLD, an excimer laser is used to ablate material from a target; atoms and ions from the resulting ablation plume of the target material are deposited on a substrate. Where necessary, a background gas may be added either as a buffer or to insure the appropriate stoichiometry on the target. The laser parameters, deposition rate, and substrate temperature govern the nucleation and growth of nanocrystals. Ag was deposited on the substrates using the PLD at room temperature to an average thickness of 1 nm. Nanoparticle formation was observed and characterized below.

RESULTS

We have successfully created two-dimensional ordered arrays of Ag nanoparticles on Si/SiO_2 using the combination of FIB milling and PLD. The FIB image Figure 1 displays a nanoscale hole array prepared in highly conductive (metallic) n-type silicon, prior to PLD deposition. The array was prepared at 80 kX magnification using a 1 pA Ga^+ ion beam with an effective spot diameter of 8 nm. Hole-depth was controlled by adjusting the total time the Ga beam dwelled at each array location, on the order of 100 msec per array site. The array in Figure 1 shows 30 nm diameter holes on 90 nm centers. This represents a limit of the current technique in the sense that collateral sputtering attributable to the wings of the Ga beam intensity distribution can significantly erode the substrate at interstitial locations between array sites for spacings less than 90 nm. This effect degrades the quality of nanoparticles arrays we are able to achieve.

Figure 1. FIB-generated secondary electron image of FIB-milled hole array. The array has 27x27 holes and measures 2.5 μm on a side. Each hole is ~30 nm diameter.

We observe different configurations of nanoparticles as a function of the depth of the hole-pattern relative to the surface of the oxide layer. Substrates prepared with shallow holes that did not penetrate the oxide layer thickness (<37nm) did not display a preference for the location of Ag islands, which formed irrespective of the presence of physical pits in the oxide surface layer, as observed by SEM and AFM observations. This result is confirmed by experiments in which FIB holes were prepared in bulk amorphous SiO_2 (glass). No preferential Ag island formation was observed in that case. This observation rules out Ga-induced Ag clustering, simple SiO defects, or SiO_2 sidewall interactions as mechanisms for the preferential nucleation at FIB damage sites.

Substrates prepared with holes that penetrated the oxide showed evidence of Ag segregation at the damage sites. The accumulated Ga dose per array site resulting in penetration of the surface oxide, and segregation of Ag to the array sites, was 2.5 x 10^6 Ga atoms, using a 4 pA beam and the appropriate dwell time. In general we observe that hole depth and apparent diameter of the holes was a function of beam dwell time, and thus total Ga dose, per array point. Deeper holes were also of a larger diameter.

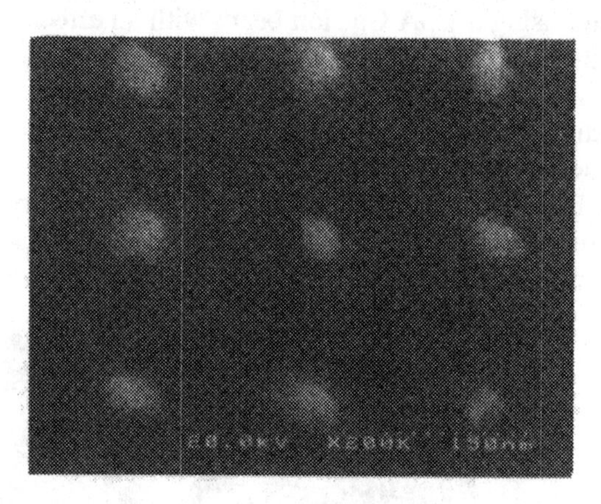

Figure 2. SEM micrograph of 40-60 nm diameter Ag nanoparticles in ordered array created by FIB and PLD. The distance between adjacent particles is 180 nm.

In the case of FIB prepared holes 30 nm or smaller in diameter, single Ag nanoparticles are observed on these selected sites as illustrated in Figure 2. In some cases, multiple particles (typically 2 per hole) have been observed in holes of slightly larger diameter.

When the FIB is used to prepare even larger diameter holes (>80 nm) on a larger array spacing (180 nm), Ag particles are observed to self-organize and decorate the periphery of the milled hole structures, as illustrated in Figure 3. We refer to these structures as "nano-bracelets." They consist of approximately 10 equi-dimensional, equi-spaced particles distributed in circular patterns centered on FIB array sites. Analysis of these structures reveals important clues to the mechanisms of Ag nanoparticle segregation in this system.

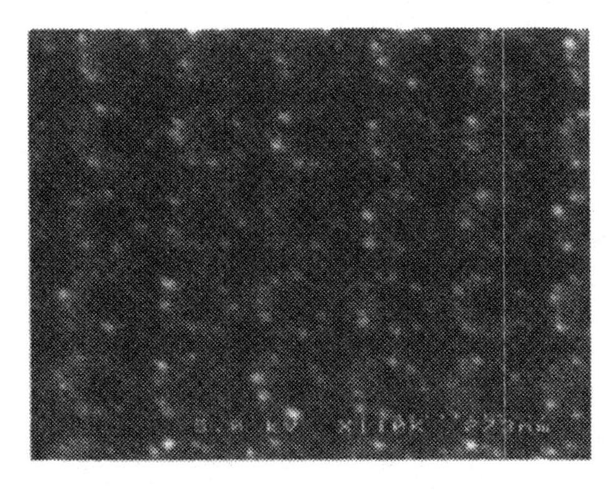

Figure 3. SEM micrograph of an ordered array of 150 nm diameter Ag "nano-bracelets" created with the FIB-PLD process. The individual particles making up the bracelets range from <10 nm to ~30 nm.

DISCUSSION

The heteroepitaxial growth of Ag on Si has been of particular interest in the literature, especially by Venables [4] and others [5, 6]. In this system interdiffusion of Ag and Si does not occur [7] resulting in sharp interfacial boundaries between the thin film deposit and the substrate. Thin film formation and growth of Ag on Si is known to proceed by a layer plus island growth Stranski-Krastinov (SK) model [7].

Of interest in this investigation is the ability of the FIB to specify the location of specialized sites with nanoscale precision. By controlling the geometry of milled structures on silicon substrates we have been able to demonstrate that the specific nucleation sites are coincident with the exposed location of the Si/SiO_2 interface. Here, we offer some ideas that may be useful in understanding the segregation of Ag clusters to predetermined sites.

In Figure 3 the oxide layer has been selectively thinned and removed in interstitial regions adjacent to milled array sites. This occurs because the Ga^+ ion beam has a presumed Gaussian distribution of intensity that extends beyond the FWHM that defines the quoted spot diameter of 8 nm. The tails of the ion intensity distribution contribute additional sputtering on the periphery of the intended damage sites. Ag nanoclusters are observed to form on this periphery, presumably coincident with the location of the interface of the Si substrate and the oxide layer, near the surface of the substrate but not at the bottom of the hole. In addition there are two other considerations. First, Ga implantation and concurrent swelling and amorphization of the implanted substrate, prior to material removal via sputtering, are known artifacts of FIB milling [1], and undoubtedly result in creation of elastic strain gradients in zones adjacent to milling sites. Second, it is generally accepted that surface and line defects such as surface steps and ledges, especially in the areas bounding milled holes that can be characterized as having surface curvature, act as preferred sites for the nucleation of metals on insulators and semiconductors [4]. Thus, there are many local mechanisms available for the defect induced nucleation of Ag clusters in the implanted regions. These mechanisms are the subject of continuing experimental investigation, and will be discussed elsewhere [8]. However, we can gain insight into the mechanism for the selection of the Si/SiO_2 interface as the preferred site for Ag segregation, as follows.

Outside the boundary of each FIB milled pattern, the Ag thin film structure on the oxide is observed to be random island growth, possibly described by a Volmer-Weber (VM) model. Clearly, either island growth model (SK or VM) is a nonequilibrium process limited by kinetic mechanisms with associated time and length scales, as has been discussed by Venables [4]. Under our experimental configuration, cluster formation occurs concurrently with PLD vapor deposition of Ag on our patterned substrates. However, following deposition, we observe that Ag segregates locally from the oxide to the FIB damage sites, with associated denuded zones surrounding each array site, containing no observable Ag clusters..

Starting with the applied Ag film thickness of 1 nm it can be shown that, in the region of the patterned substrate (Figure 3) there are available 1.9×10^6 Ag atoms per lattice point of the FIB generated array. Careful SEM analysis of the nano-bracelet structure reveals that there is an average of 10 clusters per lattice point, each averaging 20 nm in diameter. There is no evidence of substrate crystallographic preference as particles appear to be randomly distributed around the periphery of each array point. We note that this particle diameter is comparable to the smallest holes we have been able to prepare using the FIB technique. Significantly, we observe that there are no visible or detectable Ag clusters in the interstitial zones between array sites. All observable clusters reside in the bracelet pattern, as if each array point were a highly effective sink for Ag adatoms as they arrive at the substrate. If one assumes that each 20 nm diameter cluster is a hemisphere, each bracelet accounts for a total of 1.2×10^6 Ag atoms. We feel that this compares well with estimate of the 1.9×10^6 deposited adatoms available to each array point, and well within the margin of error used in this estimate.

Why should segregation occur in circular patterns associated with damaged array sites? Clearly there is axial symmetry associated with the preparation of holes using the Ga ion beam. All three defects mentioned above (the exposed Si/SiO_2 interface, implanted Ga with its associated strain energy, and steps and ledges on curved surfaces leading into the holes) can form with circular geometry.

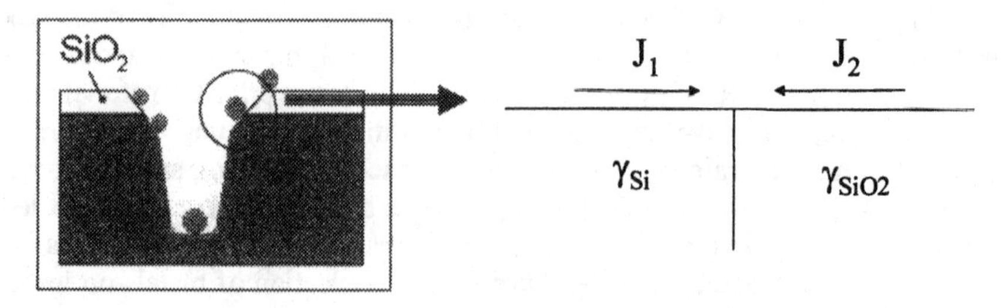

Figure 4. Various possible nucleation sites for Ag particles in the vicinity of FIB machined holes. Schematic on the right shows detail in the vicinity of the exposed Si/SiO_2 interface.

Here we will consider only the result to be expected on the basis of a difference in surface free energy and Ag diffusivity between the Si vs. the SiO_2 phases, respectively. We defer a discussion of other considerations elsewhere [8]. Consider Figure 4, which is a one dimensional model depicting the two phases separated by a sharp interface boundary. This highly schematic illustration is intended to capture the essential geometric features of the Si/SiO_2 interface as it exists on the curved sloping wall of the FIB prepared hole.

Formation of clusters is mediated, in part, through the surface diffusion of adatoms, as well as other processes. A general expression for the flux of a diffusing species based on Fick's first law can be written as

$$J_1 = -D_{Ag(Si)} \frac{\partial C_{Ag(Si)}}{\partial x} - \frac{D_{Ag(Si)} C_{Ag(Si)}}{RT} \frac{\partial E}{\partial x} \qquad [1]$$

where J_1 is the surface flux of Ag adatoms on the Si phase, $D_{Ag(Si)}$ is the surface diffusion coefficient for Ag on Si, $C_{Ag(Si)}$ is the concentration of Ag species on Si near the interface, T is the temperature, and $\frac{\partial E}{\partial x}$ is the gradient in surface free energy between the two phases across the interface. All that is required is that each phase has a different surface energy. A similar equation can be written for the flux of Ag atoms on the SiO_2 phase.

$$J_2 = -D_{Ag(SiO_2)} \frac{\partial C_{Ag(SiO_2)}}{\partial x} - \frac{D_{Ag(SiO_2)} C_{Ag(SiO_2)}}{RT} \frac{\partial E}{\partial x} \qquad [2]$$

Next we employ the concept of "local" thermodynamic equilibrium as discussed by Venables [4], applied at the junction of the free surface and the interface. Ag is uniformly applied over the entire substrate during the PLD process. So, at least in the early stages of deposition, concentration gradients do not exist, and terms of the form $\frac{\partial C}{\partial x}$ can be neglected. At equilibrium, $J_1 = J_2$, and we will determine how Ag redistributes locally in response to the gradient in surface energy that exists at the interface.

$$\frac{D_{Ag(Si)} C_{Ag(Si)}}{RT} \frac{\partial E}{\partial x} = -\frac{D_{Ag(SiO2)} C_{Ag(SiO_2)}}{RT} \frac{\partial E}{\partial x} \qquad [3]$$

Since the interface is sharp the magnitude of $\frac{\partial E}{\partial x}$ is the same whether approached from the right or the left, with the exception of a difference in sign. Canceling common factors we determine that the condition of local equilibrium is

$$\frac{D_{Ag(Si)}}{D_{Ag(SiO_2)}} = \frac{C_{Ag(SiO_2)}}{C_{Ag(Si)}} \qquad [4]$$

Here we can predict a redistribution, segregation or "pileup" of Ag at the interface, which is based on the minimization of the chemical potential of Ag as a function of position in the system. Pileup depends on a ratio of surface diffusion coefficients for Ag in each of the two phases. Note, Equation 4 has the correct limits in the case of equal mobility on each side of the boundary, i.e. we expect the concentration ratio to be 1. We expect that accumulation of Ag at the interface depends on the difference in activation energies for surface diffusion of Ag in either phase. Finally, Figure 3 clearly shows that following the accumulation of Ag at the interface, stable clusters self-organize via some form of ripening mechanism. Ongoing experiments address the kinetic processes governing this cluster formation process [8].

CONCLUSIONS

We have demonstrated that Ag deposited on oxidized Si via PLD preferentially clusters at the sites we define with the FIB. In the case of Ag nanocluster formation possible nucleation mechanisms must involve the special sites associated with the Si/SiO_2 interface. Self-organization of individual Ag clusters appears to occur on a length scale which is comparable to the smallest holes that can be prepared using the FIB. Otherwise, conglomeration of several particles can be expected at individual damage sites.

ACKNOWLEDGEMENTS

This research is supported by the U.S. Department of Energy, Office of Science, under grant DE-FG02-01ER45916.

REFERENCES

[1]C. Lehrer, L. Frey, S. Petersen and H. Ryssel, "Limitations of Focused Ion Beam Nanomachining," *Journal of Vacuum Science and Technology B,* **19**(6), 2533-2538 (2001).

[2]S. Lipp, L. Frey, G. Franz, E. Demm, S. Petersen and H. Ryssel, "Local Material Removal by Focused Ion Beam Milling and Etching," *Nuclear Instruments and Methods in Physics Research B,* **106**, 630-635 (1995).

[3]D. B. Chrisey and G. K. Hubler, Pulsed Laser Deposition of Thin Films, John Wiley & Sons, In., New York (1994).

[4]J.A. Venables, "Atomic Processes in Crystal Growth," *Surface Science,* **299/300**, 798-817 (1994).

[5]J.C. Glueckstein, M.M.R. Evans and J. Nogami, "Nanoscale Growth of Silver on Prepatterned Hydrogen-Terminated Si(001) Surfaces," *Physical Review. B,* **54**(16), R11 066 (1996).

[6]M. Sakurai, C.Thirstrup and M. Aono, "Surface Unwetting During Growth of Ag on Si(001)," *Physical Review. B,* **62**(23), 16 167 (2000).

[7]J.A. Venables, "Nucleation and Growth Processes in Thin Film Formation," *Journal of Vacuum Science and Technology*, 870-873 (1986).

[8]A.B. Hmelo, L.C. Feldman, in preparation.

Fabrication and Properties of Nanocomposites

CARBON NANOTUBES-CERAMIC COMPOSITES

E. Flahaut, S. Rul, F. Lefèvre-Schlick, Ch. Laurent and A. Peigney
CIRIMAT/LCMIE – UMR CNRS 5085
University *Paul Sabatier*
F-31062 Toulouse cedex 4, France

ABSTRACT

The use of carbon nanotubes (CNTs) as reinforcing elements in polymer-, metal- or ceramic-matrix composites is widely studied. However, the dispersion of the CNTs within the matrix is a critical step in the preparation of these composites. We have shown that a very homogeneous dispersion of the CNTs can be achieved by their synthesis *in-situ* inside an alumina-based powder. Our CCVD method produces single- and double-walled CNTs, individual or gathered in small bundles, forming a network surrounding the oxide grains. We have recently developed the synthesis of CNTs from oxides solid solution foams, prepared by the gelcasting-foam method. This preparation allows a four-fold increase in the amount of CNTs compared to the corresponding powder catalyst. Dense CNT-Fe-Al_2O_3 composites were prepared by hot-pressing as well as CNT-FeCo-$MgAl_2O_4$ and CNT-Co-MgO composites. Despite some pull-out during the fracture, a real reinforcement has not been evidenced. The CNTs provide to the composites an electrical conductivity between 0.2 and 4 $S.cm^{-1}$. We managed to align the CNTs within the ceramic by hot-extrusion, thus leading to an electrical conductivity anisotropy.

INTRODUCTION

A worldwide research effort on carbon nanotubes is currently taking place. This was initiated by Iijima's report[1] on the obtaining of carbon tubes with a diameter in the nanometer range and on their relations with fullerenes. Indeed, CNTs present some characteristics close to those of the hollow carbon fibers, which are known for several decades, their nanometric diameter and their particular structure lead to exceptional properties. A single-walled CNT (SWNT) is a perfect closed cylinder that can be visualized by rolling a graphene sheet over itself. Several concentric cylinders form a multiwalled CNT (MWNT). There are basically three main synthesis routes (arc-discharge, laser-ablation and catalytic chemical vapor deposition (CCVD)) that produce CNTs with different length, extent of bundling and defect density[2].

CNTs typically are a few nanometers in diameter and 1-100 μm in length, depending on the synthesis method. The aspect ratio is therefore very high, ranging from 10^3 to 10^5. They are extremely rigid, with a Young modulus of the order of 1 TPa, but also resist failure under repeated bending. Their tensile strength is about 50 times that of steel. The thermal conductivity is similar or higher than that of graphite and the electrical conductivity takes place with either a metallic or a semi-conducting behavior depending on the structure (helicity). CNTs are thus often considered as the ultimate carbon fibers.

Many applications are envisioned for CNTs[3], and notably much work has been carried out to incorporate them into polymer-, metal- or ceramic- matrices to form a new class of nanocomposite material[4], although most of the work published so far was dealing with CNT-polymer composites.

Several groups have prepared CNT-ceramic composites, on which we will focus here, and have studied their mechanical and electrical properties. The composites often contain SWNTs, mainly gathered into bundles (from a few to several hundreds of CNTs), or MWNTs. Ma *et al.*[5] have prepared MWNT-SiC composites (10 wt % MWNTs) by hot-pressing mixtures of CCVD MWNTs (diameter 30-40 nm) and nano-SiC powders. A densification of 94.7% was achieved by hot-pressing at 2000°C and using 1 wt % B_4C as a sintering aid. It was claimed that the presence of the MWNTs provides an increase of about 10 % of both the bending strength and fracture toughness, notably by crack deflection and MWNTs debonding. Hwang and Hwang[6] have prepared SiO_2 glass rods of micrometer size by using surfactant-MWNTs (arc-discharge) co-micelles as templates. The rods were mixed with SiO_2 and dense composites (6 wt %) MWNTs) were prepared. It was claimed that the Vickers hardness of the MWNTs/SiO_2 rods-SiO_2 materials is twice that of SiO_2 rods-SiO_2 composites but it is unclear whether this is due to the MWNTs themselves or to the change in rod-morphology they provide. DiMaio *et al.*[7] have prepared arc-discharge SWNT-SiO_2 glass composites using a sol-gel technique. The purified and shortened (50-300 nm) SWNTs were functionalized prior to ultrasonic blending in the sol phase and gelification. Heating the glasses to 600°C produced materials with 80% densification and a good optical transparency, revealing the homogenous dispersion of the SWNTs (0.1-1.0 wt%). Nonlinear optical transmission was observed. Other works will be referred to later in the present paper.

Obviously, one of the most important problems that may be encountered during the preparation of a CNT-ceramic composite is achieving a good dispersion of the CNTs. To avoid the need for any mixing, the present authors have proposed the synthesis of CNT-metal-oxide nanocomposite powders[8], containing *in situ* grown CNTs, which can be used without further processing for the preparation of dense composites.

The synthesis and the microstructure of the CNT-metal-oxide nanocomposite powders (metal: Fe, Co, Ni and their alloys; ceramic: Al_2O_3, $MgAl_2O_4$ or MgO) will be briefly discussed in the following section. Composite materials prepared by hot-pressing or hot-extrusion will then be described, as well as their mechanical and electrical characteristics.

SYNTHESIS OF CNT-METAL-OXIDE COMPOSITE POWDERS

Usually the formation of CNTs requires the presence of metal nanoparticles acting as catalyst[2]. There are two main mechanisms, the first one being often referred to as "one particle-many CNTs" and the second one, which appears to be the most active for CCVD routes, being known as "one particle-one CNT". In this case, the diameter of the CNT matches that of the corresponding particle, and it has been shown that the critical diameter below which a particle will nucleate and grow a CNT is in the range 3-5 nm[9,10]. Nanoparticles ranging from 5 to 10 nm generally end up encapsulated by graphitic shells (capsules) and nanoparticles larger than 10 nm lead to carbon nanofibers which have lost the perfect structure of CNTs and sometimes are not hollow. A key point to produce SWNTs and thin MWNTs by CCVD is thus to be able to form metal nanoparticles which are in this size range at the temperature of the formation of CNTs, which typically is 1000°C when CH_4 is used as the carbon source.

In the past, we have shown[11] that the selective reduction in H_2 of oxide solid solutions such as $\alpha\text{-}Al_{2-2x}Fe_{2x}O_3$ leads to $Fe\text{-}Al_2O_3$ nanocomposite powders containing Fe nanoparticles (< 10 nm in diameter) both inside and at the surface of the alumina grains. Interestingly, the particles start to form at a relatively high temperature (in the range 600-800°C). It was further shown[8] that the reduction in a $H_2\text{-}CH_4$ atmosphere allows to use the freshly formed surface Fe nanoparticles to form the CNTs. CH_4 decomposes on the nanoparticles and supplies them with carbon which accumulates until the saturation is reached, thus stopping the particle growth in the desired range and leading to the nucleation and growth of CNTs. The so-obtained powders (Fig. 1) contain an enormous amount of SWNTs and thin MWNTs, with a very homogenous dispersion of the CNTs which probably cannot be matched by methods involving a mechanical mixing. Incidentally, to the best of our knowledge, this was the first time that the formation of SWNTs by CCVD of a hydrocarbon was reported. Several parameters related to the starting oxide material have been investigated in order to increase the proportion of CNTs with respect to the other carbon species (carbon nanofibers, graphitic shells, disordered carbon) in the first place and then to improve the selectivity of the method towards SWNTs or double-walled CNTs (DWNTs).

These parameters include the iron content[12], the crystallographic form[13] and the specific surface area[14] of alumina-based solid solutions. The effects upon the CNT formation of some parameters related to the experimental conditions of the reduction[10,15,16,17], such as the composition of the $H_2\text{-}CH_4$ atmosphere, the reduction temperature and the time spent at that temperature have also been examined.

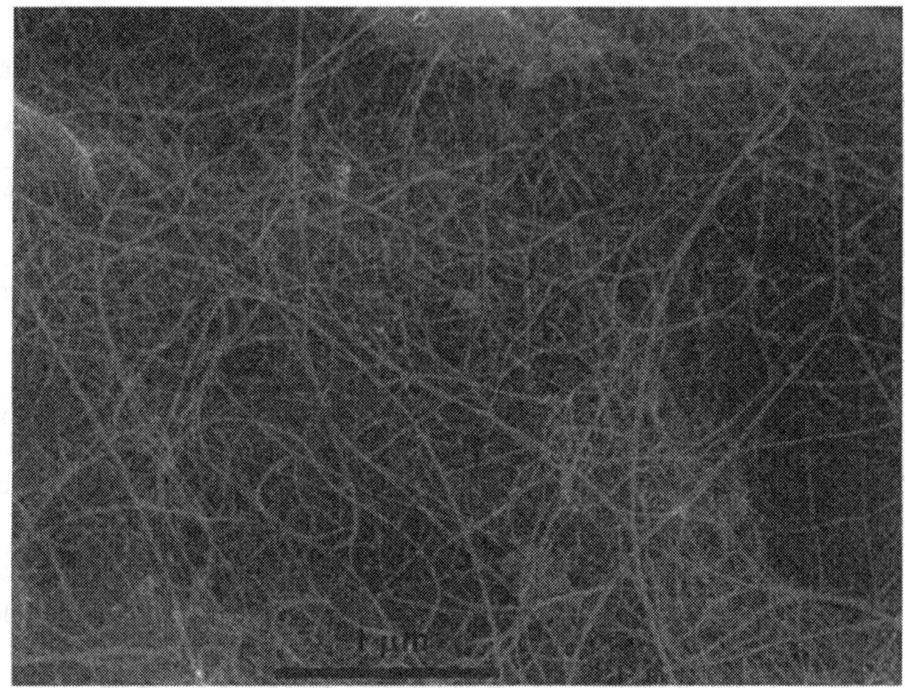

Figure 1: SEM image of CNT-Fe-Al_2O_3 nanocomposite powder.

The nature of the metallic phase (Fe, Co, Ni and their binary alloys) was also investigated[17, 18] using $MgAl_2O_4$-based solid solutions. The control of the experimental CCVD synthesis parameters using $MgAl_2O_4$-based[17, 18] and MgO-based[19] solid solutions allowed us to obtain a rather good selectivity towards the number of walls of the so-obtained CNTs, 90% of which are SWNTs and DWNTs. The CNTs are extremely well dispersed within the oxide matrix grains and their specific surface area can be as high as 790 m^2/g[20], or even higher[21]. Furthermore, it was also shown that when using MgO-based solid solutions, a simple soaking in HCl allows to separate the CNTs from the metal and MgO matrix without damaging them[19].

In order to improve both the accessibility of the metal and the diffusion of the gases by increasing the porosity, oxides foams[22, 23] have been prepared by the gelcasting-foam method. Processing the solid solution in the form of a highly (98%) porous foam allowed to multiply by 4 the amount of CNTs, because the surface accessible for CH_4 decomposition during the CCVD is greatly enhanced.

Figure 2: SEM images of a $Mg_{0.9}Co_{0.1}Al_2O_4$ foam[23] before (left) and after CCVD (right) treatment in H_2-CH_4 atmosphere.

In order to compare the CNT-metal-oxide powders with data more representative than HREM images, we have proposed a macroscopical characterization method based on the determination of the carbon quantity in the powder (C_n) and on specific surface area measurements[12, 24]. This allowed to calculate a parameter (ΔS) representing the quantity of CNTs and a parameter ($\Delta S/C_n$) representing the quality of carbon, a higher figure denoting the presence of tubes with a smaller diameter and/or CNTs with less walls and/or more carbon in a tubular form[12].

DENSE METAL-CERAMIC AND CNT-METAL-CERAMIC COMPOSITES

Microstructure and mechanical properties

In earlier studies[25-27], metal-Al_2O_3 composites have been prepared by hot-pressing the corresponding metal-Al_2O_3 nanocomposite powders[11]. The dense composites contained a very high amount of intragranular nanometric metal particles, together with sub-micronic metal particles located at the matrix grains junctions. It was shown that the presence of the intragranular metal nanoparticles provoked a transgranular fracture path, as opposed to an intergranular fracture for the unreinforced Al_2O_3, which generated a very strong increase in both fracture strength (σ_f) and fracture toughness (K_{Ic}) (Table 1). This increase was observed for small metal loading (2 wt. %, *i.e.* 1 vol. %). Other authors preparing metal-oxide nanocomposites from sol-gel powders[28] or mixtures of powders[29, 30] also reported a strong reinforcement (Table 1), although the proportion of intragranular metal particles was noticeably lower because of the routes chosen for powder preparation. Increasing the proportion of sub-micronic intergranular particles provided more reinforcement as long as the dispersion of the particles was homogeneous[26, 27, 31]. However, such a reinforcement for very low metal content was not observed in metal-$MgAl_2O_4$ composites[32], notably because a smaller oxide grain size resulted in a lower proportion of intragranular metal particles. It was thus interesting to study CNT-metal-oxide hot-pressed composites.

Table 1. Hot-pressing temperature, microstructural and mechanical data for metal-oxide composites and the corresponding oxides. Data from other research groups have been included. G_m : average size of the matrix grains; σ_f : fracture strength; K_{Ic} : SENB fracture toughness.

Specimen (wt %)	Hot-pressing temperature (°C)	Relative density (%)	G_m (µm)	σ_f (MPa)	K_{Ic} (MPa.m$^{1/2}$)	(Ref.)
Al_2O_3	1450			335	4.4	(26)
2 % Fe	1450	99.0	2	520	7.5	(27)
5 % Fe	1450	98.0	2	600	7.5	(27)
10 % Fe	1450	97.0	2	630	7.2	(27)
15 % Fe	1450	98.0	2	575	6.9	(27)
20 % Fe	1450	99.0	2	595	6.7	(27)
2 % $Fe_{0.8}Cr_{0.2}$	1450	99.0	2	690	6.5	(26)
5 % $Fe_{0.8}Cr_{0.2}$	1450	99.0	2	685	6.5	(26)
10 % $Fe_{0.8}Cr_{0.2}$	1450	98.0	2	720	6.6	(26)
15 % $Fe_{0.8}Cr_{0.2}$	1450	99.0	2	760	7.0	(26)
20 % $Fe_{0.8}Cr_{0.2}$	1450	99.0	2	860	8.0	(26)
10% Ni	> 1460	97.0	5	-	5.5	(28)
50% Ni	> 1460	74.0	5	-	8.3	(28)
11% Ni	1450	> 99.0	-	1090	4.5	(29)
12% W	1500	> 99.0	-	1105	3.7	(30)
12% Mo	1400	> 99.0	-	920	7.2	(29)
$MgAl_2O_4$	1500	96.0	1.4	259	3.3	(32)
4 % Fe	1500	98.0	0.9	289	2.3	(32)

The preparation of CNT-metal-oxide hot-pressed composites was performed as follows. The CNT-metal-oxide powders were uniaxially hot-pressed at 43 MPa in graphite dies, in a primary vacuum, at 1500°C (Al_2O_3- and $MgAl_2O_4$-matrix composites) with a dwell time fixed to 15 minutes. The dense specimens (20 mm in diameter and 2 mm thick) for mechanical tests were ground with diamond suspensions. Surfaces were polished to an optical finish and thermal and/or chemical etching treatments were adjusted to reveal the grain boundaries.

The polished surfaces, etched surfaces and fracture profiles of dense specimen were observed by scanning electron microscopy (SEM). The average grain size of the oxide (G_m) was determined by the linear intercept method. Relative densities (d %) were calculated from measurements obtained by the Archimede's method, using the density of graphite ($d_{graphite}$ = 2.25 g.cm^{-3}) for CNTs, as a first approximation.

The transverse fracture strength (σ_f) was determined by the three-point-bending test on parallelepipedic specimens (1.6 x 1.6 x 18 mm^3) machined with a diamond blade. The fracture toughness (K_{Ic}) was measured by the SENB method on similar specimens notched using a diamond blade 0.3 mm in width. The calibration factor proposed by Brown and Srawley[33] was used to calculate the SENB toughness from the experimental results. Cross-head speed was fixed at 0.1 mm/min. The values given for σ_f and K_{Ic} are the average of measures on 7 and 6 specimens, respectively.

The microstructure and mechanical properties of the hot-pressed composites have been investigated (Table 2). The presence of CNTs hinders the grains growth during hot-pressing and limits the densification to values typically around 90%. In spite of the high homogeneity of the CNTs' dispersion, the fracture strength and fracture toughness of the CNTs-containing composites are generally lower than those of the carbon-free metal-oxide composites but however comparable to those of the pure ceramics[34].

Table 2: G_m (average grain size of the oxide), d (relative density, assuming 2.25g/cm^3 - the density of graphite - for all the carbon), fracture strength (σ_f) and fracture toughness (K_{Ic}) for the hot-pressed oxides and composites.

Composite	G_m (μm)	d (%)	σ_f (Mpa)	K_{Ic}(MPa.$m^{1/2}$)	(Ref)
Al_2O_3	-	-	335	4.4	(27)
Fe-Al_2O_3	-	-	630	7.2	(27)
CNT-Fe-Al_2O_3	0.3	88.7	400	5.0	(34)
$MgAl_2O_4$	13	99.7	308	-	(34)
Fe/Co-$MgAl_2O_4$	0.8	98.2	212	2.94	(34)
CNT-Fe/Co-$MgAl_2O_4$	0.5	90.6	221	1.71	(34)

SEM observations of the fracture surfaces (Figure 3) revealed that some of the CNTs may be damaged during the hot-pressing, ending up as disordered graphitic layers which are mainly found at the matrix grain junctions, or inside pores[34]. It also showed that the fracture was mostly intergranular, which could mean that the intragranular metal nanoparticles do not contribute any more to the reinforcement. Field Emission Gun SEM (FEG-SEM) observations of fracture surfaces revealed that some CNTs seem to be trapped inside the ceramic grains, or at the grain boundaries and can be broken during the fracture with some degree of pull-out which can contribute to the absorption of some fracture energy[34]. However, no reinforcement is apparent at the macroscopic scale, and it is assumed that this is due mainly to the poor densification. The control of the interface between the CNTs and the matrix is also one of the key issues to achieve some mechanical reinforcement.

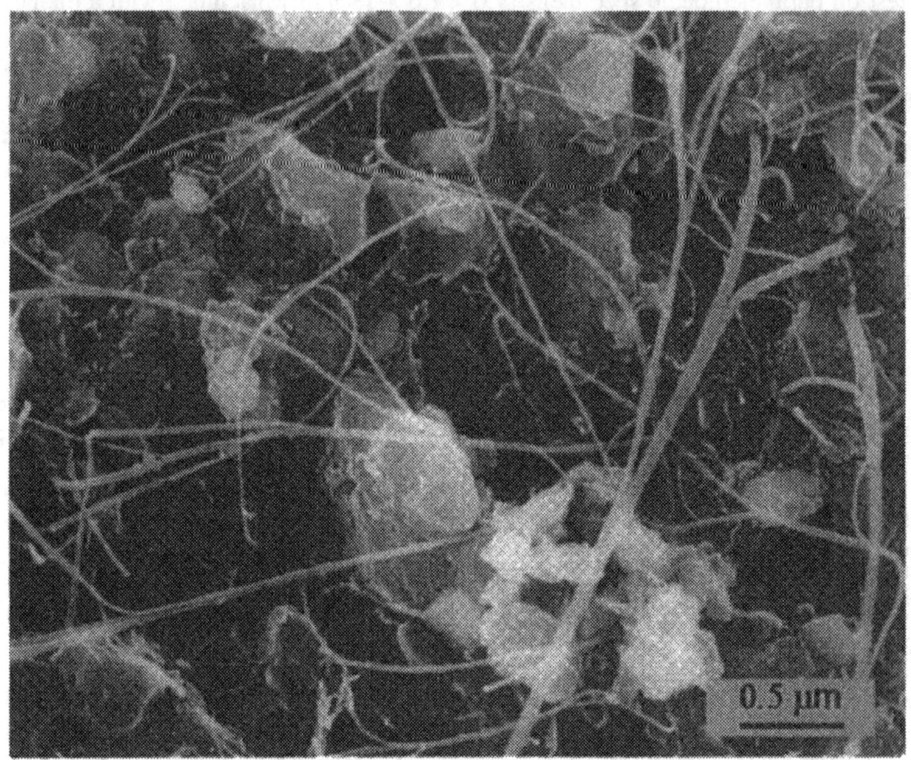

Figure 3: SEM fracture image of a CNT-Fe-Al_2O_3 hot-pressed composite.

Indeed, these results must be compared with those of Zhan *et al.*[35] who have prepared almost fully densified SWNTs-Al_2O_3 composites by sintering ball-milled mixtures of nanocrystalline alumina and SWNTs at a temperature as low as 1150°C using the Spark Plasma Sintering (SPS) technique. Because the temperature and the time needed are much lower than in conventional hot-pressing, no damage of the CNTs was observed. SEM images show that the bundles of CNTs are mainly located at the grains boundaries and that the contact with alumina is good and also that the fracture is intergranular. These authors report a very large gain in fracture toughness (from Vickers indentation) for the composites containing 10 vol. % SWNTs (from 3.7 $MPa.m^{1/2}$ for pure Al_2O_3 to 9.7 $MPa.m^{1/2}$). Sun *et al.*[36] have reported a colloidal process to coat CNTs with alumina prior to mixing with a concentrated alumina suspension, so that the final material contains only 0.1 wt. % CNTs. SPS was used to densify the composite at 1300°C. The Vickers hardness is slightly higher than that of pure alumina (17.6 GPa *vs* 16.9 GPa) and the Vickers indentation-derived fracture toughness is increased from 3.7 to 4.9 $MPa.m^{1/2}$ with only 0.1 wt. % CNTs. It is proposed that coating the CNTs before sintering results in an improved bonding with the matrix. Crack-bridging and CNTs pull-out could also lead to the increase in fracture toughness. These studies[35-36] have thus demonstrated that CNT-ceramic composites warrant further investigation with respect to the mechanical properties[37].

Electrical properties

The electrical conductivity of the oxides (Al_2O_3, $MgAl_2O_4$), metal-oxide composites and CNT-metal-oxide composites was investigated. The electrical conductivity of the hot-pressed specimens was measured at room temperature with DC currents along the length of parallelepipedic specimens (1.6 x 1.6 x 8 mm^3), *i.e.* perpendicularly to the hot-pressing axis. The current densities used were lower than 160 mA/cm^2. The first measures indicated that the values of the electrical conductivity were closely correlated to the amount of CNTs within the composites and conductivities ranging from 0.2 to 4 $S.cm^{-1}$ were measured.

It was attempted, for the first time with a ceramic matrix, to align the CNTs by hot-extrusion[38]. Dense billets were prepared by high-temperature extrusion in a graphite die, under a primary vacuum. The die profile was manufactured from the design proposed by Kellett *et al.*[39] for the extrusion of ZrO_2-based ceramics. The large (14 mm diameter) and small (6.7 mm) cylinders are joined by a conical part with a cone angle of 26°. Lubrication was accomplished by folding sheets of graphite paper over the various surfaces of the die. Firstly, the composites were partially densified under a mild load (8.6 MPa) and then the materials were extruded under a 43 MPa load. Al_2O_3 and $MgAl_2O_4$-matrix materials were extruded at 1500°C. Cylindrical rods, 6.7 mm in diameter and 20 mm long were obtained.

The alignment of the CNTs could be confirmed by an anisotropy of the electrical conductivity between the direction parallel to the extrusion direction and perpendicular to the latter ($\sigma_{//}/\sigma_{\perp}$ = 30). The conductivity measured along the extrusion direction (20 $S.cm^{-1}$) was noticeably higher than for hot-pressed composites. Work is in progress to determine the percolation threshold for CNT-metal-oxide nanocomposites and the results will be published elsewhere.

CONCLUSIONS

CNT-metal-oxide powders (Al_2O_3, $MgAl_2O_4$) are prepared by reduction in H_2-CH_4 of the corresponding oxide solid solutions. These composite powders contain a very high amount of SWNTs and DWNTs, very homogenously dispersed. The CNTs are very long and tend to form small bundles. Using solid solution in the form of porous foams, as opposed to powders, allows a four-fold increase of the quantity of CNTs. The CNT-metal-oxide composite materials prepared by hot-pressing are electrical conductors (0.2 - 4 $S.cm^{-1}$) whereas the corresponding oxide and metal-oxide materials are insulating. Alignment of the CNTs by hot-extrusion results in an anisotropy of the electrical conductivity. The densification of the CNT-containing composites is relatively poor (less than 90%) and therefore no mechanical reinforcement was observed. However, CNTs mechanically blocked between the matrix grains or those which are entrapped inside the matrix grains could contribute to the reinforcement on a microscopical scale. Other groups using the SPS technique to fully densify the composites have indeed reported a rather strong increase in the Vickers indentation-derived fracture toughness. More studies are necessary to investigate the fracture strength, build on these preliminary yet promising results.

REFERENCES

[1]S. Iijima, "Helical microtubules of graphitic carbon", *Nature*, **354** 56-58 (1991).

[2]Ch. Laurent, E. Flahaut, A. Peigney and A. Rousset, "Metal nanoparticles for the catalytic synthesis of carbon nanotubes", *New Journal of Chemistry*, 1229-1237 (1998).

[3]E. G. Rakov, "The chemistry and applications of carbon nanotubes", *Russian Chemical Reviews*, **70** [10] 827-863 (2001).

[4]Ch. Laurent and A. Peigney, "Carbon nanotubes in composite materials", *Encyclopedia of Nanotechnology*, H. S. Nalwa Ed., Academic Press, accepted (2003).

[5]R. Z. Ma, J. Wu, B. Q. Wei, J. Liang and D. H. Wu, "Processing and properties of carbon nanotube/nano-SiC ceramic", *Journal of Materials Science,* **33** [21] 5243-5246 (1998).

[6]G. L. Hwang and K. C. Hwang, "Carbon nanotube reinforced ceramics", *Journal of Materials Chemistry,* **11** [6] 1722-1725 (2001).

[7]J. DiMaio, S. Rhyne, Z. Yang, K. Fu, R. Czerw, J. Xu, S. Webster, Y. P. Sun, D. L. Carroll and J. Ballato, "Transparent silica glasses containing single-walled carbon nanotubes", *Information Sciences*, **149** [1-3] 69-73 (2003).

[8]A. Peigney, Ch. Laurent, F. Dobigeon and A. Rousset, "Carbon nanotubes grown in situ by a novel catalytic method", *Journal of Materials Research*, **12** [3], 613-615 (1997).

[9]J. H. Hafner, M. J. Bronikowski, B. R. Azamian, P. Nikolaev, A. G. Rinzler, D. T. Colbert, K. A. Smith and R. E. Smalley, "Catalytic growth of single-wall carbon nanotubes from metal particles", *Chemical Physics Letters*, **296** 195-202 (1998).

[10]A. Peigney, P. Coquay, E. Flahaut, E. De Grave, R. E. Vandenberghe and Ch. Laurent, "A study of the formation of single- and double-walled carbon nanotubes by a CVD", *Journal of Physical Chemistry B*, **105** [40] 9699-9710 (2001).

[11]X. Devaux, Ch. Laurent and A. Rousset, "Chemical synthesis of metal nanoparticles dispersed in alumina", *Nanostructured Materials*, **2** [4] 339-346 (1993).

[12]A. Peigney, Ch. Laurent, O. Dumortier and A. Rousset, "Carbon nanotubes-Fe-Alumina nanocomposites. Part I: influence of the Fe content on the synthesis of powders", *Journal of the European Ceramic Society*, **18** [14] 1995-2004 (1998).

[13]Ch. Laurent, A. Peigney and A. Rousset, "Synthesis of carbon nanotube-Fe-Al_2O_3 nanocomposite powders by selective reduction of different $Al_{1.8}Fe_{0.2}O_3$ solid solutions", *Journal of Materials Chemistry*, **8** [5] 1263-1272 (1998).

[14]Ch. Laurent, A. Peigney, E. Flahaut and A. Rousset, "Synthesis of carbon nanotubes-Fe-Al_2O_3 powders : influence of the characteristics of the starting $Al_{1.8}Fe_{0.2}O_3$ oxide solid solution", *Materials Research Bulletin*, **35**, [5] 661-673 (2000).

[15]A. Peigney, Ch. Laurent and A. Rousset, "Influence of the composition of a H_2-CH_4 gas mixture on the catalytic synthesis of carbon nanotubes-Fe/Fe_3C-Al_2O_3 nanocomposite powders", *Journal of Materials Chemistry*, **9** [5] 1167-1177 (1999).

[16]E. Flahaut, A. Peigney and Ch. Laurent, "Double-walled carbon nanotubes in composite powders", *Journal of Nanoscience and Nanotechnology*, **3** 151-158 (2003).

[17]A. Govindaraj, E. Flahaut, Ch. Laurent, A. Peigney, A. Rousset and C. N. R. Rao, "An investigation of carbon nanotubes obtained from the decomposition of methane over reduced $Mg_{1-x}M_xAl_2O_4$ (M = Fe, Co, Ni) spinel catalysts", *Journal of Materials Research*, **14** [6] 2567-2576 (1999).

[18]E. Flahaut, A. Govindaraj, A. Peigney, Ch. Laurent, A. Rousset and C. N. R. Rao, "Synthesis of single-walled carbon nanotubes using binary (Fe, Co, Ni) alloy nanoparticles prepared *in situ* by the reduction of oxide solid solutions", *Chemical Physics Letters*, **300** [1-2] 236-242 (1999).

[19]E. Flahaut, A. Peigney, Ch. Laurent and A. Rousset, "Synthesis of single-walled carbon nanotube-Co-MgO composite powders and extraction of the nanotubes", *Journal of Materials Chemistry*, **10** [2] 249-252 (2000).

[20]R. R. Bacsa, Ch. Laurent, A. Peigney, W.S. Bacsa, Th. Vaugien and A. Rousset, "High specific surface area carbon nanotubes from catalytic chemical vapor deposition process", *Chemical Physics Letters*, **323** 566-571 (2000).

[21]E. Flahaut, R. Bacsa, A. Peigney and Ch. Laurent, " Gram-Scale CCVD Synthesis of Double-Walled Carbon Nanotubes", *Chemical Communications*, accepted (2003).

[22]S. Rul, Ch. Laurent, A. Peigney and A. Rousset, "Carbon nanotubes prepared in situ in a cellular ceramic by the gelcasting-foam method", *Journal of the European Ceramic Society*, **23** [8] 1233-1241 (2003).

[23]S. Rul, *D. Phil Thesis, University Paul Sabatier, Toulouse, France* 185p (2002).

[24]A. Peigney, Ch. Laurent, E. Flahaut, R. Bacsa and A. Rousset, "Specific surface area of carbon nanotubes and bundles of carbon nanotubes", *Carbon*, **39** 507-514 (2000).

[25]A. Rousset, X. Devaux, M. Brieu and A. Marchand, "Ceramic-metal nanocomposites", Proc. "*10th Tsukuba General Symposium*", Tsukuba (Japan), **134** (1990).

[26]X. Devaux, Ch. Laurent, M. Brieu and A. Rousset, "Microstructural and mechanical properties of ceramic matrix nanocomposites", *Comptes rendus de l'Académie des Sciences, Paris*, **312** [12] 1425-1430 (1991).

[27]X. Devaux, Ch. Laurent, M. Brieu and A. Rousset, "Ceramic-matrix nanocomposites", "*Composites Materials*", ed. by A. T. Di Benedetto, L. Nicolais and R. Watanabe, Elsevier Science Publishers B. V., Amsterdam, 209-214 (1992).

[28]E. Breval, Z. Deng, S. Chiou and C. G. Pantano, "Sol-gel prepared nickel-alumina composite materials. Part I. Microstructure and mechanical properties", *Journal of Materials Science*, **27** [6] 1464-1468 (1992).

[29]K. Niihara, A. Nakahira and T. Sekino, "New nanocomposite structural ceramics", *Materials Research Society Symposium Proceedings*, **286** 405-412 (1993).

[30]M. Nawa, T. Sekino and K. Niihara, "Fabrication and mechanical behavior of Al_2O_3/Mo nanocomposites", *Journal of Materials Science*, **29** [12] 3185-3192 (1994).

[31]Ch. Laurent and A. Rousset, "Metal-oxide matrix ceramic nanocomposites", *Key Eng. Mater.*, **108-110** (1995), 405-421.

[32]O. Quénard, Ch. Laurent, M. Brieu and A. Rousset, "Synthesis, microstructure and oxidation of Co-MgAl2O4 and Ni-MgAl2O4 nanocomposite powders", *Nanostructured Materials*, **7** [5] 497-507 (1996).

[33]W. F. Brown and J. E. Srawley, "Plane strain crack toughness testing of high strength metallic materials", ASTM Special Technical Publication, **410**, ASTM, Philadelphia, PA, USA, (1966).

[34]E. Flahaut, A. Peigney, Ch. Laurent, Ch. Marlière, F. Chastel and A. Rousset, "Carbon nanotube-metal-oxide nanocomposites : microstructure, electrical conductivity and mechanical properties", *Acta Materialia*, **48** [14] 3803-3812 (2000).

[35]G. D. Zhan, J. D. Kuntz, J. Wan and A. K. Mukherjee, "Single-wall carbon nanotubes as attractive toughening agents in alumina-based Nanocomposites", *Nature Materials*, **2** [1] 38-42 (2003).

[36]J. Sun, L. Gao and W. Li, "Colloidal processing of carbon nanotube/alumina composites", *Chemistry of Materials*, **14** [12] 5169-5172 (2002).

[37]A. Peigney, "Tougher ceramics with nanotubes", *Nature Materials*, **2** [1] 15-16 (2003).

[38]A. Peigney, E. Flahaut, Ch. Laurent, F. Chastel and A. Rousset, "Aligned carbon nanotubes in ceramic-matrix nanocomposites prepared by high-temperature extrusion", *Chemical Physics Letters*, **352** [1-2] 20-25 (2002).

[39] B. J. Kellet, C. Carry and A. Mocellin, "High-temperature extrusion behavior of a superplastic zirconia-based ceramic", *Journal of the American Ceramic Society*, **73** [7] 1922-1927 (1990).

PLASMA REACTION SYNTHESIS OF ALUMINA-ALUMINIUM OXYNITRIDE NANOCOMPOSITE POWDERS

Sreeram Balasubramanian*, Rajendra K. Sadangi*, Vijay Shukla[†], Bernard H. Kear* and Dale E. Niesz*
*Department of Ceramic and Materials Engineering
[†]Department of Mechanical and Aerospace Engineering
Rutgers University, Piscataway, New Jersey, USA

ABSTRACT

Fully dense nanocomposite materials with grain size sufficiently smaller than the wavelength of light will not scatter light at grain boundaries. Such materials have potential for use as transparent armor materials with excellent optical and mechanical properties. In these materials, significant retardation to grain growth during densification is achieved when the volume fractions of the constituent phases are nearly equal. Commercially available Al_2O_3 and AlN powders were agglomerated by spray-drying, and then subjected to plasma melting and quenching. This yielded a mixture of transparent powder that was comprised of a homogeneous Al_2O_3-rich extended solid solution phase and some translucent/opaque powder that was comprised of unreacted α-Al_2O_3 particles (that scattered light) in the Al_2O_3-rich extended solid solution phase. More homogeneous melting and easier formation of the metastable solid solution were achieved when using sub-micron size Al_2O_3 particles, but AlN particle size did not have a significant effect. An addition of MgO also promoted formation of the solid solution, possibly by forming a lower temperature eutectic during plasma melting. Upon heat treatment in nitrogen atmosphere, α-Al_2O_3 and AlON phases nucleated from the Al_2O_3-rich solid solution. This powder was transparent even after soaking at 1450 C for 10 h, thereby suggesting that the α-Al_2O_3 and AlON grains were of a nanocrystalline size. The melt-quenching process is a viable method to synthesize AlON-base composites from Al_2O_3 and AlN powders.

INTRODUCTION

Currently, thick laminates of soda lime glass and polycarbonate are being used as transparent armor materials. Thinner layers of light-weight, ballistically harder materials will provide significant weight and thickness savings compared to current materials. Single-crystal sapphire and polycrystalline materials like aluminum oxynitride (AlON) and magnesium aluminate spinel ($MgAl_2O_4$) are some of the suitable alternative materials [1].

Single crystal materials are expensive and often difficult to fabricate into complex shapes. The optical quality of polycrystalline materials is reduced by scattering of light by grain boundary impurities and pores [1]. Also non-cubic materials like Al_2O_3 scatter light at grain boundaries due to anisotropy in refractive index [1]. Fabrication of high-purity, fully-dense polycrystalline materials with clean grain boundaries typically requires high temperatures and large pressures. This typically results in materials with large grain sizes. However, the mechanical properties of materials decrease with increasing grain size.

The presence of second phase inclusions increases the diffusion length for grain growth by zener effect [2, 3]. Such retardation of grain growth is maximized when near equal volume fractions of multiple phases are present in a material. Kim et al. obtained significantly smaller grain sizes by increasing the volume fraction of Al_2O_3 and $MgAl_2O_4$ phases in an Al_2O_3-$MgAl_2O_4$ -ZrO_2 composite material [2, 3].

Composite compositions with large grain sizes normally produce too much optical scatter, due to difference in refractive index of constituent phases, to be used as optical materials. However, if the grain size of the constituent phases is sufficiently smaller (less than 1/10) than the wavelength of light to be transmitted, no scattering occurs at the grain boundaries and composite-optical materials with excellent optical and mechanical properties might be possible [1].

AlON is a solid solution of Al_2O_3 and AlN [4]. It may be regarded as a nitrogen-stabilized cubic γ-Al_2O_3 phase [5]. The optical, mechanical and chemical properties of AlON have been widely investigated [4, 6-9]. Synthesis methods for AlON under different conditions have been investigated [5, 10-17]. The stable phase region of AlON is limited [18]. Hence, control of oxygen and nitrogen potentials is critical while forming AlON and excess oxidation or nitridation should be avoided [5, 15, 17].

Plasma spraying is a widely used technique to obtain ceramic coatings from oxides such as alumina, titania, chromia and silica [19]. It is also a versatile technique that can be universally used to obtain new materials comprising non-equilibrium phases that are homogeneously mixed at the nano-scale. These non equilibrium phases nucleate and grow into equilibrium phases on heat treatment; hence, such homogeneous precursor powders obtained by plasma melt-quenching can be used to fabricate homogeneously mixed nanocomposites. Recently, plasma melt quenching has been used to process Al_2O_3-base ceramics [17, 20-22]. The process is characterized by rapid quenching after melting at high temperatures. Due to the short reaction times corresponding to the short residence time (on order of milliseconds) of the powder in the plasma flame, the synthesis reactions are complete by the time the powder exits the plasma flame. Hence, a special atmosphere is not necessary, and the resulting powder can be collected in water after passing through air. This rapid reaction rate and ability to work without an inert atmosphere allows plasma melt-quenching to be a rapid reaction technique that can be scaled for the production of AlON-base materials.

In a prior investigation, we reported our initial results on the feasibility of synthesis of a metastable Al_2O_3-rich, Al_2O_3-AlON solid solution phase by plasma melt quenching in water [23]. In this paper, results on controlled nucleation of nanocrystalline Al_2O_3 and AlON phases by heat treatment of the above metastable powder are presented. These plasma melt quenched powders were transparent even after heat treatment at temperatures up to 1450°C. This suggests that suggesting that they are comprised of nanocrystalline grains of Al_2O_3 and AlON. These results indicate that plasma melt-quenching is a feasible technique to synthesize precursor powders that can be processed to obtain dense, bulk transparent AlON-base composites.

EXPERIMENTAL METHOD

Commercially available Al_2O_3 (91 wt. %) and AlN (9 wt. %) powders were dispersed with Duramax D-3005 (Rohm & Haas Co., Philadelphia, PA) and ball milled as an aqueous system. Dilute nitric acid was added to lower the pH to the acidic range and prevent hydrolysis of AlN [24, 25]. Compositions of the batches used in this investigation are listed in Table I. Some of the Al_2O_3 powders used in this study contained 0.5 wt. % MgO. The resulting slurries were spray dried to obtain uniformly mixed, free flowing feed material for plasma melt-quenching. The spray dried feed powder was heated in flowing argon at ~1200 °C for 2 hours. This sintering treatment was necessary to provide bonding between particles, thus preventing particle disintegration in the plasma flame.

Table I. Composition of the different powders

Batch	Average particle size (μm) Al_2O_3	Average particle size (μm) AlN	% MgO (wt. %)
A	1.4*	3***	0.0
B	0.2**	3	0.0
C	0.2	3	0.5
D	0.2	1***	0.0
E	0.2	1	0.5

*A-152 GR, Alcoa Inc., Pittsburgh, PA
**RC-UFX, BMI -BAIKOWSKI Malakoff Inc., Malakoff, TX
***Dow Chemical Co., Midland, MI (3 μm = XUS 35569 and 1 μm = XUR-YM-2002-97923)

The powders were side injected into an Ar/10% H_2 plasma flame, using a Metco 9MB torch (Sulzer Metco, Westbury, NY). After exiting the plasma flame, they were passed through air and collected in a water-bath. The resulting powder was dried and classified into different sizes. These were then heat treated to temperatures up to 1450°C.

RESULTS AND DISCUSSION

The constituent powders melt and react inside the plasma flame. The completeness of melting of the feed powder depends primarily on the powder melting point, residence time, flame temperature, particle size distribution in the starting powder, agglomerated feed powder size, density, and thermal conductivity. In general, at least two passes through the plasma flame were necessary to ensure that a major fraction of the feed powder particles experience melting and homogenization prior to water quenching.

Figure 1 shows XRD patterns of powders obtained by melt-quenching different sizes of aggregated feed powder of batches B and C. These batches differed only in MgO content. As indicated, the plasma melt quenched powder is composed of γ-Al_2O_3 (nitrogen-stabilized cubic Al_2O_3) and unreacted α-Al_2O_3 and AlN phases. The amount of γ-Al_2O_3 increases with decreasing feed size. The finest feed powder gives primarily γ-Al_2O_3 phase (Al_2O_3-AlON solid solution) and much less unreacted Al_2O_3 and AlN phases. In contrast, the largest feed powder has a significant percentage of unreacted Al_2O_3 and AlN particles.

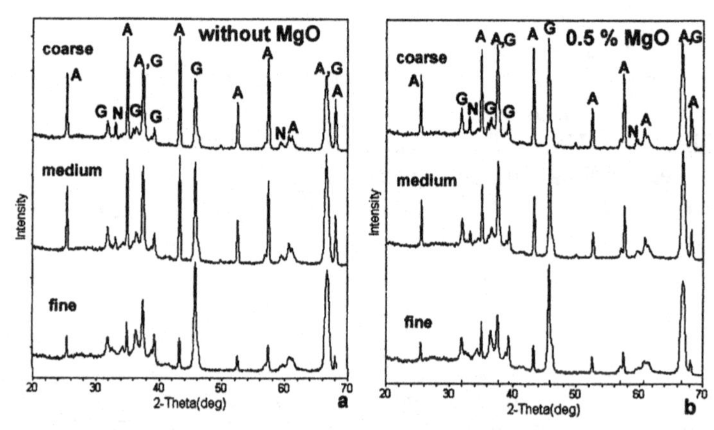

Figure 1. XRD patterns of plasma melt-quenched powders obtained from feed powder with different sizes: (a) without MgO (Batch B) and (b) with 0.5 wt. % MgO (Batch C); **A**-α-Al_2O_3, **G**- γ-Al_2O_3 (Al_2O_3-AlON solid solution), **N**- AlN

For similar aggregate size, the amount of the γ-Al_2O_3 phase (extended solid solution) relative to the unreacted α-Al_2O_3 phase is more for the powders containing MgO. Apparently, this is because MgO forms a eutectic that facilitates melting of the agglomerated particles.

Finer Al_2O_3 particles in the agglomerated feed powder aided the formation of γ-Al_2O_3 solid solution, but AlN particle size did not have any significant effect. These results were presented elsewhere [23].

Figure 2 shows optical micrographs of the different sized plasma melt quenched powders of Batch C, corresponding to the XRD pattern in Figure 1b. The fine plasma melt-quenched powders are comprised almost entirely of transparent particles. The amount of translucent/opaque particles increases with increasing size. These results indicate that the non-transparency of some of the plasma-melt-quenched powders is due to the scattering of light by the unreacted

α-Al_2O_3 particles. Some of large powders have melted completely and appear transparent. These possibly had a longer residence time in the plasma flame and, hence, attained higher temperatures and reacted completely to form the γ-Al_2O_3 phase (Al_2O_3-AlON solid solution).

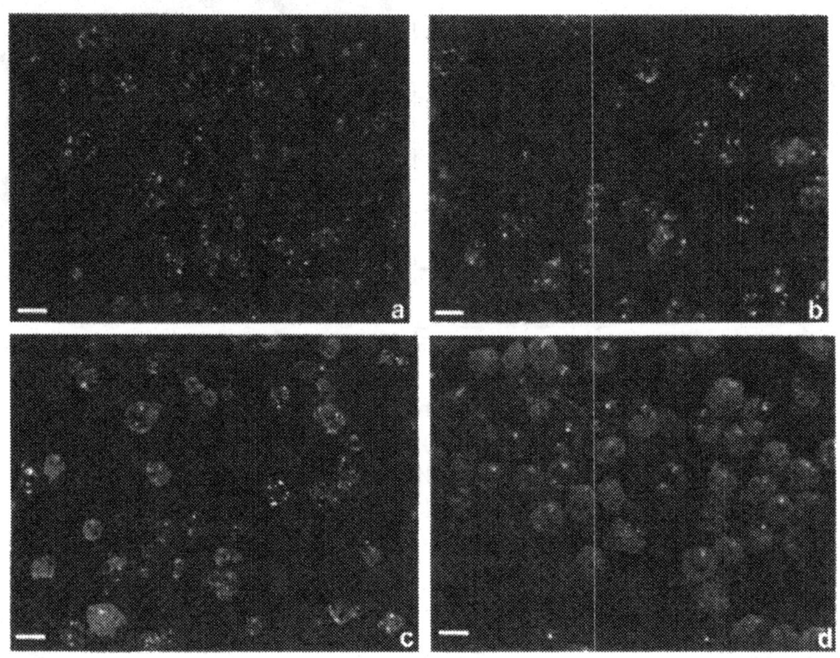

Figure 2. Optical micrographs of plasma melt quenched powders of Batch C: fine size-a, b; medium size- c and coarse size- d. (bar = 20 μm in a, c, d; =10 μm in b)

This demonstrates the limited capacity of the current experimental set-up to melt and homogenize large feed particles. However, such large feed powder particles can be melted completely by increasing their residence time in the plasma flame. To accomplish this, an inductively coupled plasma with axial powder feed is being developed.

The melt-quenched powder that contained mostly γ-Al_2O_3 phase (Batch C, < 38 μm) was heat-treated at temperatures up to 1450°C for several hours in a nitrogen atmosphere. AlON and α-Al_2O_3 phases start to nucleate from the γ-Al_2O_3 solid solution at 1200°C, Figure 3. However, the amount of α-Al_2O_3 is much less than the γ-Al_2O_3 up to 1300°C. At 1400°C, co-nucleation of α-Al_2O_3 and AlON phases is nearly complete, and the amounts of α-Al_2O_3 and AlON phases are comparable. Heat-treatment at 1450°C for 10 hours does not cause any significant change in the relative amounts of the constituent phases. Figure 4 shows optical micrographs of this powder (Batch C, fine size, 1450°C- 10 h). The powder is transparent indicating there is no significant scattering despite the mismatch in refractive indices of the constituent Al_2O_3 and AlON phases. This suggests that the α-Al_2O_3 and AlON grains are significantly smaller than the wavelength of light.

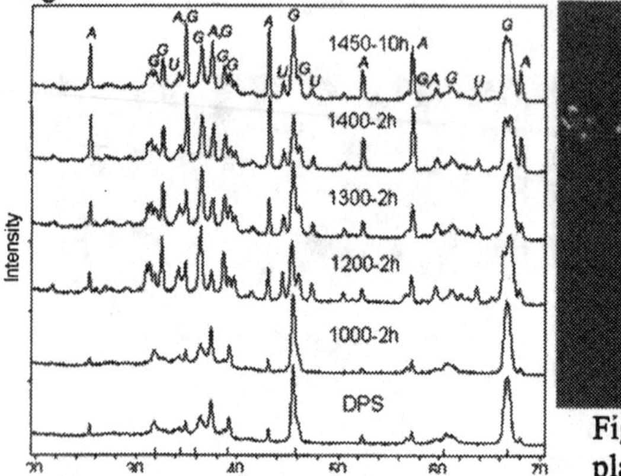

Figure 3. XRD patterns of plasma melt-quenched powder heat treated to different temperatures. Starting powder (Batch C) was finer than 38 μm and contained 0.5 wt. % MgO; **A**-α-Al_2O_3, **G**- γ-Al_2O_3 (Al_2O_3-AlON solid solution), **U**- unknown

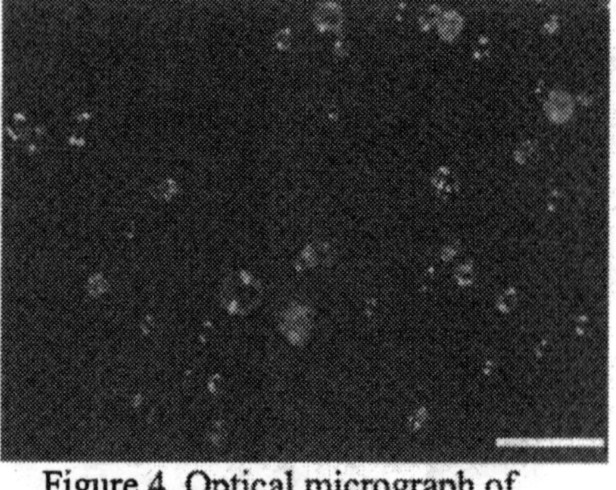

Figure 4. Optical micrograph of plasma melt quenched powders of Batch C, heat treated 1450°C- 10 h; bar = 50 μm.

These results indicate that the precursor powder obtained by plasma melt-quenching can be used to fabricate an AlON-base composite. The homogeneous plasma melt-quenched powder can be reduced to a finer size by mechanical attrition, and the resulting fine particles can be densified by pressure-assisted sintering at ~1400°C to obtain a fine nanocomposite structure.

SUMMARY AND CONCLUSIONS

Commercially available Al_2O_3 and AlN powders were agglomerated by spray-drying, and then subjected to plasma-melt-quenching in an Ar-H_2 atmosphere. This yielded a mixture of transparent powder that was comprised of a homogeneous Al_2O_3-rich extended solid solution phase and some translucent/opaque particles that was comprised of unreacted α-Al_2O_3 particles (that scattered light) in the Al_2O_3-rich extended solid solution phase. The formation of this solid solution phase was facilitated by using fine feed powders with sub-micron size Al_2O_3 particles containing MgO. AlON and α-Al_2O_3 phases co-nucleated upon heat-treatment of the precursor powder in a nitrogen atmosphere. The nucleation was complete by 1400°C and the powder remained transparent even after soaking for 10 h at 1450°C. These results indicate that a precursor powder obtained by plasma melt quenching in water can be used to fabricate AlON-base composites.

ACKNOWLEDGEMENT

Research was sponsored by the Army Research laboratory [ARMAC-RTP] and was accomplished under the ARMAC-RTP Cooperative Agreement Number DAAD19-01-2-0004. The views and conclusions contained in this document are those of the authors and should not be interpreted as representing the official policies, either expressed or implied, of the Army Research Laboratory or the U.S. Government. The U.S. Government is authorized to reproduce and distribute reprints for Government purposes notwithstanding any copyright notation hereon.

REFERENCES

[1]D. C. Harris, Materials for Infrared Windows and Domes, SPIE Press, Bellingham WA, 1999.

[2]B. –N. Kim, K. Hiraga, K. Morita, and Y. Sakka, " Superplasticity in Alumina Enhanced by co-dispersion of 10% Zirconia and 10% Spinel Particles," *Acta Mater.*, **49**, 887-95 (2001)

[3]B. –N. Kim, K. Hiraga, K. Morita, and Y. Sakka, " A High-Strain-Rate Superplastic Ceramic," *Nature (London)*, **413**, [6853] 288-91 (2001)

[4]N.D. Corbin, "Aluminium Oxynitride Spinel: A Review," *J. Eur. Ceram. Soc.*, **5** 143-154 (1989).

[5]J. Zheng and B. Forslund, "Carbothermal Synthesis of Aluminium Oxynitride (ALON) Powder: Influence of Starting Materials and Synthesis Parameters," *J. Eur. Ceram. Soc.*, **15**, 1087-100 (1995).

[6]T.M. Hartnett, S.D. Bernstein, E.A Maguire, R.W. Tustison, "Optical Properties of ALON (aluminum oxynitride)," *Infrared Physics & Tech.*, **39** [4] 203-211 (1998).

[7]H.X. Willems, P.F. Van Hal, G. de With, and R. Metselaar, "Mechanical Properties of γ-Aluminium Oxynitride," *J. Mater. Sci*, **28** 6185-89 (1993).

[8]H.X. Willems, P.F. Van Hal, R. Metselaar, and G. de With, "AC-Conductivity Measurements on γ-Aluminium Oxynitride," *J. Eur. Ceram. Soc.* **15** [11] 57-61 (1995).

[9]P. Goeuriot, D. Goeuriot-Launay, and F. Thevenot, "Oxidation of an Al_2O_3-γAlON Ceramic Composite,"*J. Mater. Sci.*, **25** 654-60 (1990).

[10]L. Yawei, L. Nan, and Y. Runzhang, "Carbothermal Reduction Synthesis of Aluminium Oxynitride Spinel Powders at Low Temperatures," *J. Mater. Sci. Lett.*, **16** 185-86 (1997).

[11]J.W. McCauley and N.D. Corbin, "Phase Relations and Reaction Sintering of Transparent Cubic Aluminium Oxynitride Spinel,"*J. Am. Ceram. Soc.*, **62** [9] 476-79 (1979).

[12]I. Adams, T.R. AuCoin, and G.A. Wolff, "Luminescence in the System Al_2O_3-AlN," *J. Electrochem. Soc.*, **109** [11] 1050-54 (1962).

[13]V.J. Silvestri, E.A. Irene, S. Zirinsky, and J.D. Kuptis, "Chemical Vapor Deposition of $Al_xO_yN_z$ Films," *J. Electron. Mater.*, **4** 429-44 (1975).

[14]M. Ish-Shalom, "Formation of Aluminium Oxynitride by Carbothermal Reduction of Aluminium Oxide in Nitrogen," *J. Mater. Sci. Lett.*, **1** 147 (1982).

[15]W. Rafaniello and I.B. Cutler, "Preparation of Sinterable Cubic Aluminium Oxynitride by Carbothermal Nitridation of Aluminium Oxide," *J. Am. Ceram. Soc.*, **64** C128 (1981).

[16]L. Yawei, L. Nan, and Y. Runzhang, "The Formation and Stability of γ-Aluminium Oxynitride Spinel in the Carbothermal Reduction and Reaction Sintering Process," *J. Mater. Sci.*, **32** 979-82 (1997).

[17]H. Fukuyama, Y. Nakao, M. Susa, and K. Nagatta, "New Synthetic Method of Forming Aluminium Oxynitride by Plasma Arc Melting," *J. Am. Ceram. Soc.*, **82** [6] 1381-87 (1999).

[18]H X. Willems, M.M.R.M. Hendrix, G. de With, and R. Metselaar, "Thermodynamics of Alon II: Phase Relations," *J. Eur. Ceram. Soc.*, **10** 339-46 (1992).

[19]B. Kear, L.E. Cross, J.E. Keem, R.W. Siegel, F.A. Spaepen, K.C. Taylor, E.L. Thomas and K.N. Tu, "Research Opportunities for Materials With Ultrafine Microstructure," National Materials Advisory Board, National Academy Press, Washington, DC, 1989.

[20]L.H. Cao, K.A. Khor, L. Fu and F. Boey, "Plasma Processing of Al_2O_3/AlN Composite Powders," *J. Mater. Proc. Tech.*, **89-90** 329-98 (1999).

[21]B H. Kear, Z. Kalman, R K. Sadangi, G. Skandan, J. Colaizzi and W E. Mayo, "Plasma-Sprayed Nanostructured Al_2O_3/TiO_2 Powders and Coatings," *J. of Th. Spray Tech.* **9** [4] 483-487 (2000).

[22]J. Colaizzi, W.E. Mayo, B.H. Kear, S.-C. Liao, "Dense Nanoscale Single- and Multi-Phase Ceramics Sintered by Transformation Assisted Consolidation," *Intn'l J. Powder Metall.*, **37** [1] 45-54 (2001).

[23]S. Balasubramanian, R. K. Sadangi, V. Shukla, B. H. Kear and D. E. Niesz, "Plasma Melt Quenching of AlON Ceramics for Armor Applications," In Press, Proceedings of Cocoa Beach Conference of Am. Ceram. Soc., FL, 2002.

[24]S. Novak and T. Kosmac, "Interactions in Aqueous Al_2O_3-AlN Suspensions During the HAS Process," *Mater. Sci. Eng.*, **A256** 237-42 (1998).

[25]P. Bowen, J.G. Highfield, A. Mocellin and T.A. Ring, "Degradation of Aluminium Nitride in an Aqueous Environment," *J. Am. Ceram. Soc.*, **73** [3] 724-28 (1990).

NANOPHASE DECOMPOSITION IN PLASMA SPRAYED $ZrO_2(Y_2O_3)/Al_2O_3$ COATINGS

Fude Liu, Frederic Cosandey, XinZhang Zhou and Bernard H. Kear
Dept. of Ceramic and Materials Engineering
Rutgers University
Piscataway, NJ 08854-8065

ABSTRACT

The microstructure and nanophase decomposition of metastable $ZrO_2(3Y_2O_3)/Al_2O_3$ coatings formed by air plasma spraying was analyzed by combined TEM techniques of nanoprobe diffraction and X-Ray EDS spectroscopy. For the $ZrO_2(3Y_2O_3)/20\%Al_2O_3$ composition, the initial splat-quenched metastable structure consists of a high volume fraction of primary ZrO_2-rich particles with a thin intergranular phase of eutectic composition. Nanoprobe diffraction analysis revealed the presence of two phases: cubic-ZrO_2, and intergranular nanoscale δ-Al_2O_3. Upon heat treatment at 1200°C, the tetragonal ZrO_2 phase appears with a fine-intergranular dispersion of nanoscale α-Al_2O_3. When the annealing temperature reaches 1400°C, the microstructure evolves into an equilibrium non-segregated nanocomposite structure consisting the α-Al_2O_3 grain dispersed within the ZrO_2 matrix.

INTRODUCTION

Yttria-stabilized ZrO_2, is widely used as a coating material, because of its good thermal shock resistance and low thermal conductivity, making it the material of choice for thermal barrier coatings [1]. It is also used in situations where its high strength and toughness can be exploited, for example where wear resistance is important [2]. Over the years, several methods have been developed for depositing ZrO_2-base coatings on substrate surfaces, including sol-gel, magnetron sputtering [3], electron-beam evaporation [4], and plasma spraying [5]. In plasma spraying, the coating material, usually in a powder form, is injected into a plasma stream, where it is melted and then quenched on a substrate surface to form a coating layer. Dense nano-composite ceramic coatings are produced when a significant fraction of the feed particles undergo melting before quenching on the substrate surface. The characteristics of the coating are dependent on a number of variables, including particle size, distribution and morphology, particle temperature and velocity, torch/substrate stand-off distance, and deposition rate. The technique has the advantages of simplicity and versatility, and hence has been widely used [6]. In a previous study [7], the influence of plasma melting and quenching on the structure of $ZrO_2(3Y_2O_3)/20Al_2O_3$ particles was studied by X-Ray diffraction (XRD) and field emission scanning electron microscopy (FESEM). In this previous investigation it was shown that the effect of rapid quenching and solidification of the molten particles upon impact with a chilled metallic substrate was to develop a duplex structure, consisting of a uniform distribution of primary ZrO_2 particles in a eutectic-like matrix phase. Moreover, it was found that the scale of the duplex structure was dependent on the solidification rate, with the finest structures corresponding to the fastest cooling rates. In this study we expand our analysis by performing an in depth characterization analysis using analytical TEM observations, including bright field image, selected area diffraction, nano-probe diffraction and X-Ray EDS analysis, to provide new insights into the microstructure of the splat-quenched and annealed coating structure.

EXPERIMENTAL PROCEDURES

A Sultzer-Metco arc-plasma torch was used for plasma spraying. The working gas was argon/10%H_2 at 75 psi; the carrier gas was argon at 60 psi. Further details can be found elsewhere [7]. Low-carbon steel coupons (30 x 5 x 2 mm thick), prepared by grit blasting and acetone cleaning, were used as substrate materials. In a typical experiment, a coupon was attached to a rotating wheel and plasma sprayed until a coating thickness of about 3 mm was developed. After deposition, the coating was removed from its steel substrate by dissolving the steel in nitric acid.

Thin films for TEM examination were prepared first by cutting the coating into samples 2.4 x 2.4 x 2 mm thick using a low-speed saw. Each sample was taper-polished by diamond polishing using the "Tripod" method, and then further thinned using a Gatan dual-ion miller. Both in-plane and transverse samples were prepared for TEM examination. TEM observations were made at 200 kV using a Topcon 002B microscope. Bright field images were used to study the grain size and texture, while selected area electron diffraction and nano-probe diffraction were used to determine crystal structure. The composition of the phases was determined by X-Ray EDS spectroscopy using a PGT detector and analysis system. For quantitative chemical analysis of Y, Zr and Al content, the Cliff-Lorimer k factors were used while for oxygen, stoichiometry to the metal atoms was assumed.

RESULTS AND DISCUSSION

It is envisioned that drastic undercooling of the melt occurs during rapid solidification, thereby inducing prolific nucleation of primary ZrO_2-rich particles in the liquid phase. Thus, a transient semi-solid or mushy zone is developed, wherein the primary particle size is of sub-micron scale dimensions. At the same time, due to segregation, the composition of the liquid phase shifts towards the eutectic, thus forming an Al_2O_3-rich matrix phase. An interesting consequence of

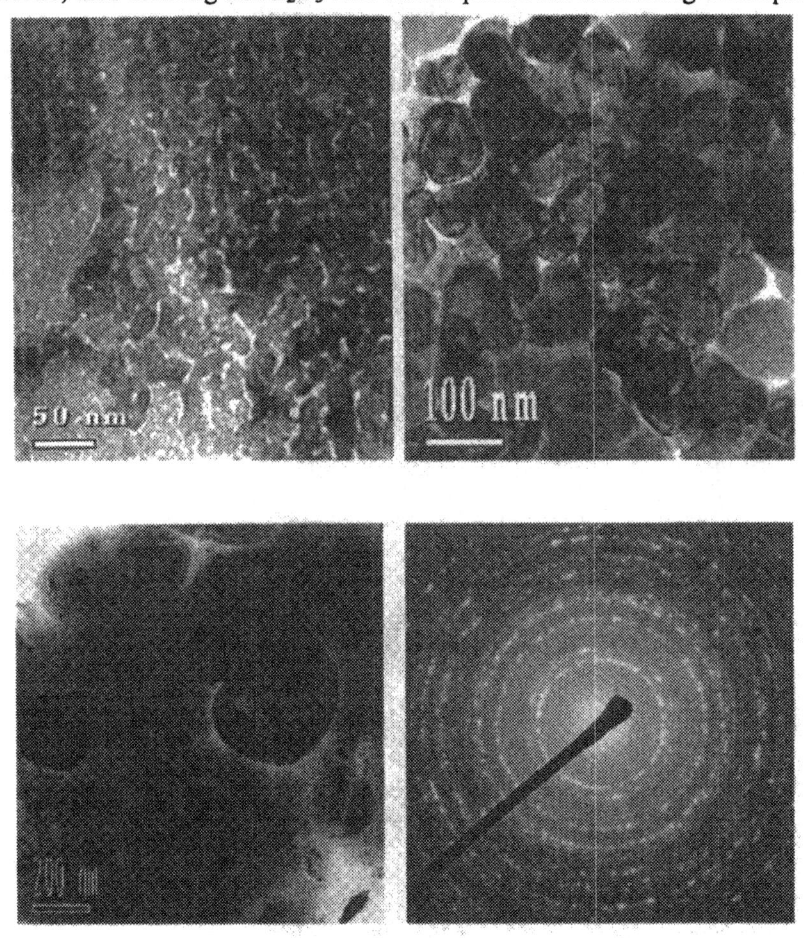

this effect is the ability to generate nano- to micro-scale composite structures by controlled thermal decomposition of the initial metastable material.

As-Deposited Material – The microstructure varies from area to area with representative bright-field TEM images of an in-plane section, i.e. taken parallel to the coating surface and within 200 μm of the steel substrate, are shown in Fig.1a-c, with a representative selected-area diffraction (SAED) pattern shown in Fig.1d. The micrograph shows a relatively high volume fraction of primary ZrO_2 particles with a size that varies from area to area ranging from 20 to 400 nm with a mean size of about 50 nm. This is an indication that the cooling rate of the splat-cooled liquid droplets is not uniform. In between the cellular ZrO_2 particles a thin intergranular film can be observed. It has been estimated that the cooling rates experienced by a typical splat-quenched particle, about 1-2 micron in thickness, is ~10^6K/sec [7]. When a coating is formed by superposition of splats, the effective cooling rate diminishes with increasing coating thickness, due to the thermal resistance of the previously splat-quenched material. This reaches a steady state value after about 10-20 deposited layers, provided that precautions are taken to continuously cool the substrate during deposition, as was the case here. Even in an optimal situation, however, the cooling rate experienced by different particles, due to variations in particle size, velocity and temperature, will vary from one location to the next during coating build-up, which accounts for the range of primary particle size observed, cf. Fig.1 (a),(b) and (c).

From the SAED pattern of Fig.1d, the most prominent diffraction rings can be best indexed as cubic-ZrO_2 as summarized in Table I .

Table I. List of observed d spacing taken from the SAED in Fig 1d and indexing of the reflections corresponding to cubic ZrO_2 and metastable δ-Al_2O_3.

Measured d (nm)	Cubic-ZrO_2 d (nm)	δ - Al_2O_3 d (nm)
0.298	0.2939 (111)	
0.257	0.2545 (200)	
0.239	-	0.2386 (213)
0.212	-	0.2165 (207)
0.198	-	0.1979 (220)
0.181	0.1799 (220)	
0.154	0.1535 (311)	
0.147	0.1469 (222)	
0.140	-	0.1411 (327)
0.128	0.1272 (400)	
0.118	0.1168 (313)	
0.114	0.1138 (420)	

However, we note here that it is difficult to distinguish between cubic and tetragonal phases in this system, because of peak overlap due to strain and segregation effects. Our previous work by XRD however [7], has shown that both phases are present in the splat-quenched material, but that the cubic phase predominates at the highest cooling rates. The additional reflections can be attributed to the metastable tetragonal δ form of Al_2O_3. Further information on the metastable phase identification by nano-probe diffraction is presented below.

It is interesting to note that the orientation of the individual grains of this cellular structure depicted in Fig. 1 is not random but is strongly textured. This is shown in Fig. 2b where a selected area pattern from a 0.6-μm area is depicted. The SAED pattern revealed that within this area, all the primary ZrO_2-rich particles have a common orientation with the same [111] zone axis.

The composition and structure of the intergranular phase as well as those of the primary particles have been determined by X-ray EDS and nano-probe diffraction and the results are shown in Fig. 3. The X-ray EDS analysis shows that the primary particles are ZrO_2-rich and the intergranular phase is Al_2O_3 rich with a composition approaching the Al_2O_3-ZrO_2 eutectic composition. In other words, both phases are highly supersaturated relative to the equilibrium state, due to the rapid solidification experienced by the splat-quenched material.

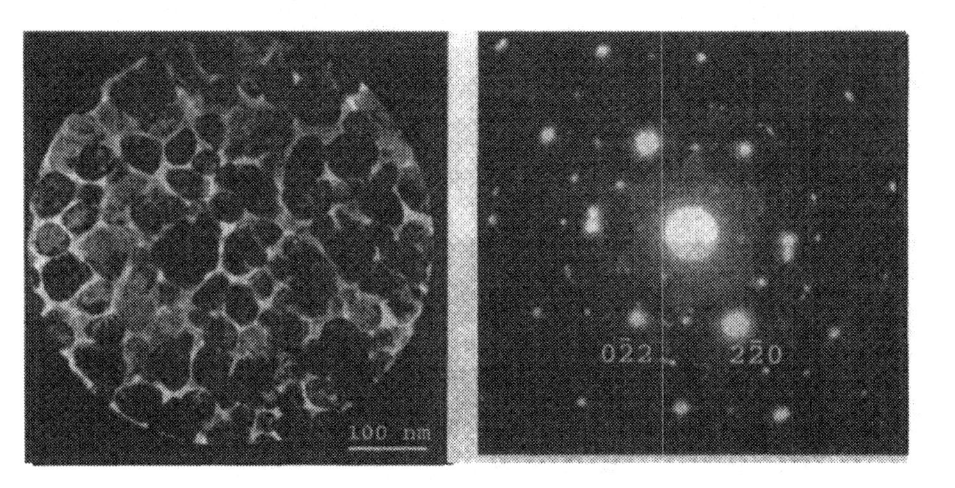

From nano-probe diffraction shown in Fig. 3c, the primary particles are identified as cubic ZrO_2 in a [011] zone axis orientation while the intergranular phases is identified from Fig. 3d as tetragonal δ-Al_2O_3, in a [5,18,2] zone axis orientation, which, according to Figure 3b, is also supersaturated with ZrO_2. This nano-probe diffraction pattern could also be indexed according to the cubic γ-Al_2O_3 crystal structure albeit with a worse fit. Possibly, the stain associated with the presence of ZrO_2 in extended (supersaturated) solid solution could stabilize the metastable alumina phases and preclude unique crystal structure identification. It is interesting to note that the presence of any Al_2O_3 phase was not detected by XRD [7]. This is probably due to the extremely small grain size and weak X-Ray scattering of Al_2O_3 phases.

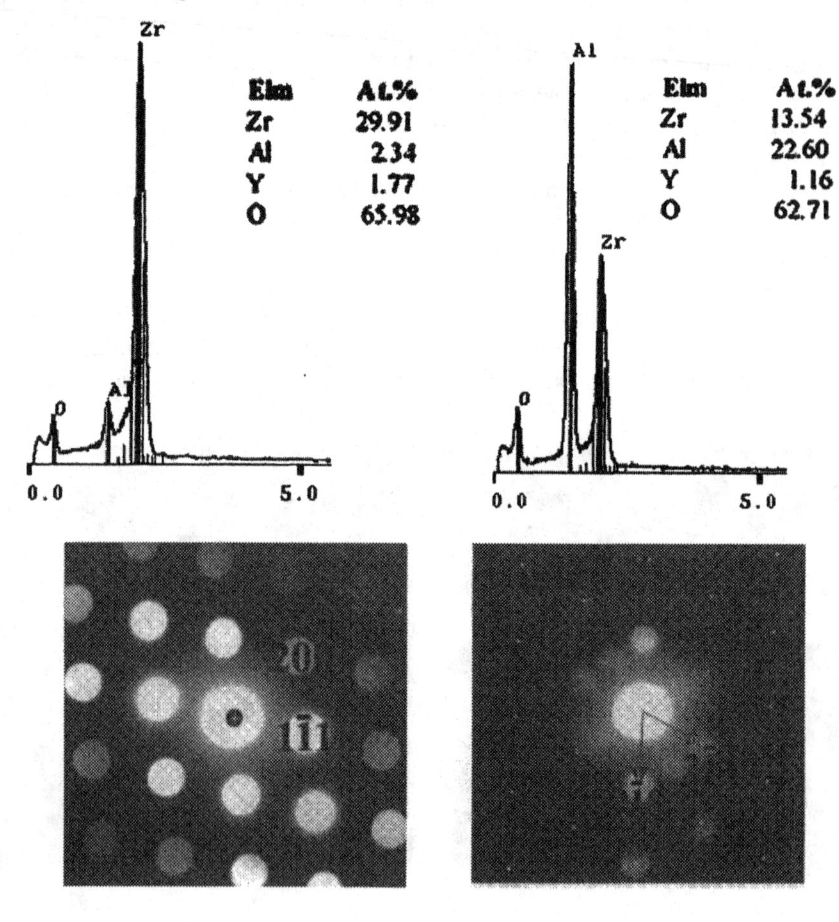

When viewed in transverse sections, some of the primary particles are elongated, apparently in the direction perpendicular to the coating surface as shown in Fig. 4.

This suggests that the primary particle growth in the splat-quenched material is controlled by the steep temperature gradient. Thus, it is envisioned that primary particles nucleated in the undercooled liquid become elongated during growth in a direction perpendicular to the rapidly advancing solid-liquid interface.

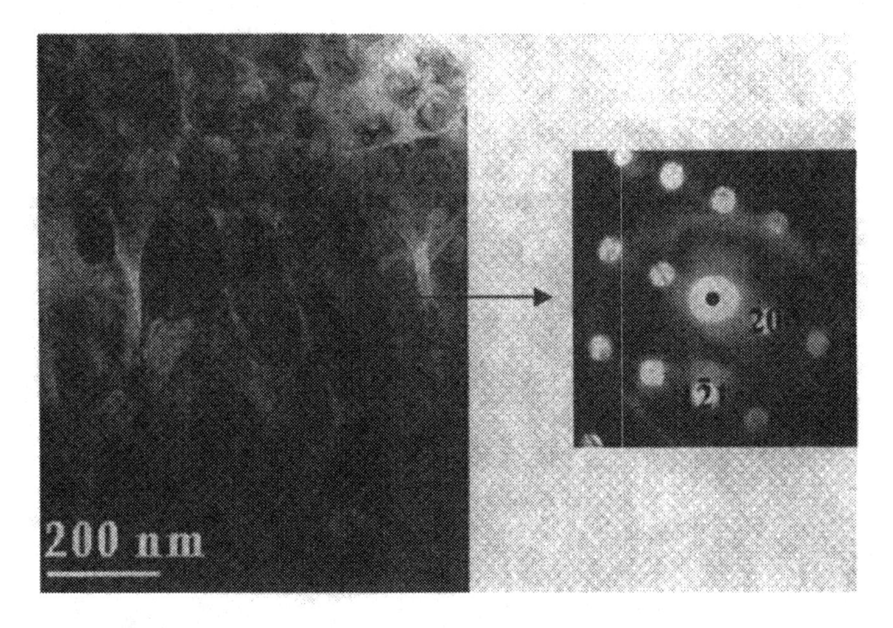

Taking this model one step further, it is expected that with a smaller driving force for solidification, i.e. a lower cooling rate, the primary particles nucleated in the mushy zone will develop into continuous fibers in the growth direction, hence forming a textured cellular structure. A cellular structure has been seen in water-quenched particles, where the effective cooling rate is much lower than that of splat-quenched material. It remains to be seen if the cellular structure itself degenerates into a dendritic structure at even lower cooling rates, as would be expected. In all cases examined so far, nucleation and growth of primary particles in the undercooled liquid is accompanied by segregation, which is more pronounced with decreasing cooling rate, or in other words with general coarsening of the microstructure.

Heat-Treated Material – Since the evidence described above demonstrates that the rapidly solidified (splat-quenched) material has a far-from-equilibrium or metastable structure, upon heat treatment the material should decompose into its corresponding equilibrium two-phase structure (Al_2O_3 + ZrO_2), as dictated by the phase diagram. Previous observations by combined SEM and XRD showed this to be the case [7]. Moreover, it was established that the granularity of the duplex structure can be controlled from nano- to micro-scale dimensions via heat treatment, as would be expected.

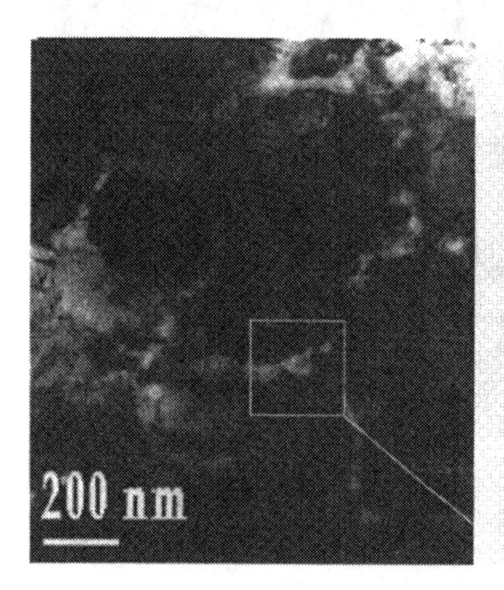

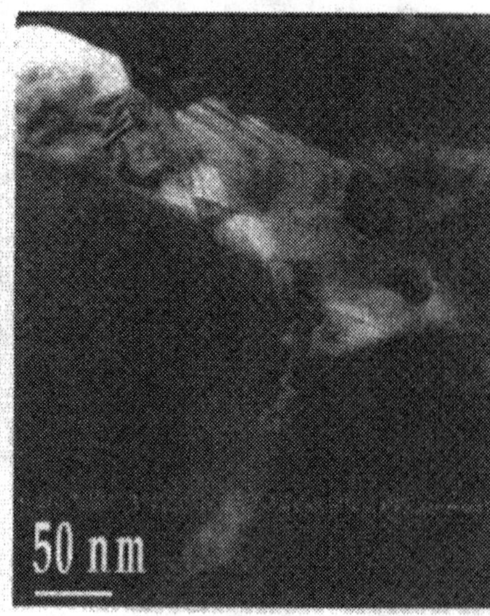

Heat Treated Material at 1200°C. Figure 5 shows a typical TEM bright-field image (in-plane section) of the splat-quenched coating material, after heat treatment at 1200°C. Once again, there is clear evidence from nano-probe diffraction analysis for the formation of a duplex structure of α-Al_2O_3 + tetragonal-ZrO_2. Moreover, it appears that the formation of particles of the dispersed α-Al_2O_3 phase is accompanied by some recrystallization of the primary ZrO_2 phase. In addition, from Fig. 5b, it can be seen clearly that the nucleation and growth of α-Al_2O_3 particles occurs preferentially in the segregated interphase region of the original cellular structure.

Heat Treated Material at 1400°C. After heat treatment at 1400°C, c.f. Fig. 6, the equilibrium duplex structure is clearly resolved into its constituent α-Al_2O_3 + tetragonal ZrO_2 phases. These two phases were identified by nano-probe diffraction as shown in Fig. 6b for tetragonal ZrO_2 in a [-2,-2,1] zone axis orientation and in Fig. 6c for α-Al_2O_3 in a [4,-1,2] zone axis orientation. More careful examination of Fig. 6 reveals that the ZrO_2 matrix phase contains a high density of dislocations, whereas the α-Al_2O_3 dispersed phase is essentially dislocation free. The effect of residual stress can be ascribed to thermal expansion mismatch stresses, arising during cool-down from the heat-treatment temperature. The effect places the α-Al_2O_3 particles in compression and the surrounding ZrO_2 matrix phase in tension. Since α-Al_2O_3 is much harder than tetragonal ZrO_2, the absence of plastic deformation in the dispersed phase is understandable. Any plastic accommodation during the cool-down cycle necessarily must occur in the softer matrix phase.

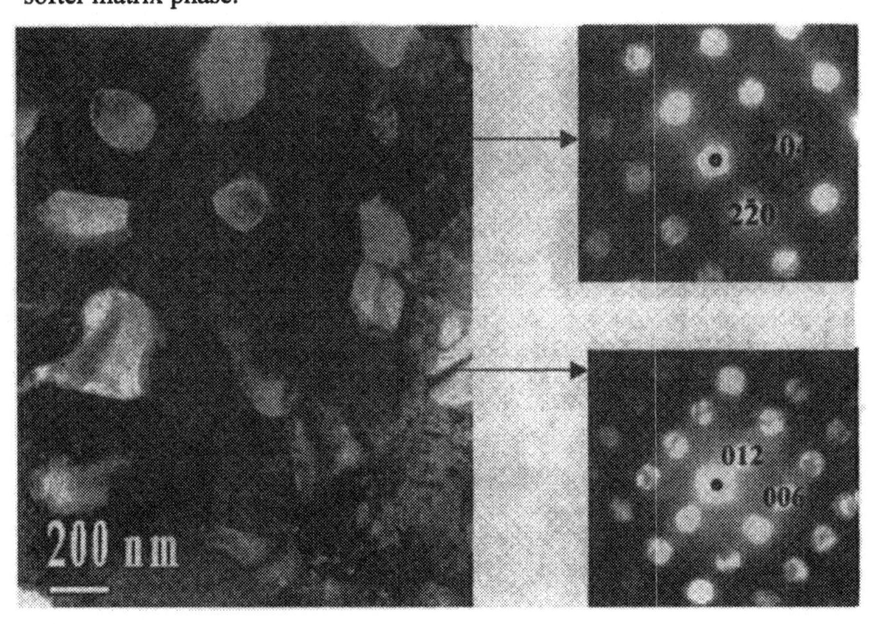

The monolinic-ZrO_2 has never been found in any of the samples studied. Presumably, this is because the addition of 3 mole-% Y_2O_3 is sufficient to stabilize the tetragonal-ZrO_2 phase. However, we note that Y_2O_3 is distributed between both phases in the as-deposited material, c.f. Fig. 3, which raises the question as to its partitioning between α-Al_2O_3 and tetragonal ZrO_2 phases in the final equilibrium structure. Further work is needed to resolve this question.

CONCLUSIONS

1. Nanocomposite ceramic $ZrO_2(3Y_2O_3)/20Al_2O_3$ coatings were formed by rapid quenching on low-carbon steel substrates.
2. After quenching, TEM observations reveal a duplex microstructure, consisting of a high volume fraction of primary cubic-ZrO_2 particles with a nanocrystalline δ-Al_2O_3-rich intergranular phase. For ZrO_2, the particle size ranged from about 20 to 400 nm with a mean size of about 50 nm.
3. Upon final heat treatment at 1400 °C, an equilibrium duplex microstructure is formed consisting of a fine distribution of α-Al_2O_3 particles dispersed within a tetragonal-ZrO_2 matrix.

REFERENCES

1. T. Sato, S. Ohtaki, and M Shimada, "*Transformation of Yttria Partially Stabilized Zirconia by Low Temperature Annealing in Air*," J. Mater. Sci., **20**, 1466-1470 (1985).
2. J. D. Lin and J. G. Duh, "*The use of x-ray line profile analysis in the tetragonal to monoclinic phase transformation of ball milled, as-sintered and thermally aged zirconia powders*," J. Mater. Sci., **32** (12), 4901-4908 (1997).
3. Z. Ji, J. A. Haynes, M. K. Ferber, and J. M. Rigsbee, "*Metastable tetragonal zirconia formation and transformation in reactively sputter deposited zirconia coatings*," Surface and Coatings Technology, **135**, 109-117 (2001).
4. J. F Groves, "*Directed Vapor Deposition*," Ph.D, University of Virginia, 1998.
5. R. Suryanarayanan, *Plasma Spraying: Theory and Application* (World Scientific, Singapore, 1993).
6. H. -S. Ahn, J. -Y. Kim, and D. -S. Lim, "*Tribological behaviour of plasma-sprayed zirconia coatings*," Wear, **203-204**, 97 (1997).
7. X. Zhou, V. Shukla, R.W. Cannon, and B.H. Kear, "*Metastable Phase Formation in Plasma Sprayed Zirconia (Yttria)-Alumina Powders*," J. Amer. Ceram. Soc., **In Press** (2003).

RAMAN IMAGE OF THE SiC FIBERS NANOSTRUCTURE

M. Havel
LADIR-UMR 7075
CNRS & Université P. & M. Curie,
2 rue Henry Dunant, 94320 Thiais, France;
ONERA, DMSC,
29 avenue de la division Leclerc,
92320 Châtillon, France

Ph. Colomban
LADIR-UMR 7075
CNRS & Université P. & M. Curie,
2 rue Henry Dunant, 94320 Thiais
France

ABSTRACT

Unique properties (conductivity, diffusion, reactivity, sintering, mechanical strength...) have been reported for nanostructured materials, all of which result from the interfacial characteristics. Raman microspectrometry appears to be the only efficient and non-destructive analysis method allowing the imaging of the crystalline or amorphous nanophases in these materials. This technique provides information on the nature, size and strain of the nanophases. When used in resonant conditions, it allows probing a fiber's surface specifically (a few tens nanometers). Obviously, a comparison is possible with Scanning Electron Microscopy, but an optical analysis has many advantages since it is remote and needs little or no sample preparation. In this paper, we take SiC fibers, which are a heterogeneous system of nanophases, as examples. The effects of chemical and thermal aging treatments are observed on first generation NLM-Nicalon® and last generation Tyranno SA3® fibers. α and β SiC have been identified on 1600°C annealed NLM® fiber's surface. Raman mapping of NLM® fiber section gave evidence that carbon nanosized moieties show homogeneous concentration, nature and size (~ 0.4 nm) across the fiber. In contrast, carbon moieties analyzed on Tyranno SA3® fiber section show different concentration, nature, size (< 3 nm) and reactivity (regarding molten $NaNO_3$ action) on fiber's core and skin which are attributed to the elaboration process.

INTRODUCTION

Materials made with (or from) nanophases received considerable attention in the last few years but their characterisation is not easy. Usual X-Ray diffraction techniques fail to analyse nanosized/nanocrystalline materials: small size effect and short-range disorder are hardly distinguished and long counting times are required. In contrast, Raman spectroscopy has demonstrated its capability to analyse low crystallinity and amorphous materials. Most of the properties controlled by the particle size (electrical conductivity, mechanical strength, etc.) are correlated to Raman parameters, which can therefore be used to predict these properties [1]. This can help to understand SiC fibers' aging mechanisms in view of their mechanical and chemical resistance optimisation.

In this paper, we focus on the SiC fiber's nanophases identification and on their specific distribution in the fibers. In order to point out the most important aging factors, we selected fibers from first and last (3rd) generations to highlight the specific behaviour of each phase regarding aging.

NLM-Nicalon® fiber (Nippon Carbon, Japan) was the first commercially available fiber made from polymer precursor. Polycarbosilane (PCS) reticulation is obtained by thermal oxidation of Si-H and C-H bonds, leading to the formation of Si-O-Si and Si-O-C bridges. As a result, the oxygen content is high, reaching more than 10% wt. It has been shown that this fiber contains 56% wt of SiC nanocristals and 10% wt of free turbostratic carbon, which are linked in an intergranular $SiO_xC_{1-x/2}$ (x~0.12) oxicarbide phase [2]. The estimated residual porosity is about 2% and the fiber's surface is protected by a silica film [3].

Tyranno SA3® fiber (UBE Industry, Japan) is synthesized from graft PCS where methyl groups are replaced by organic groups containing aluminium as a crystal growth inhibitor. Reticulation is obtained by 160°C thermal oxidation, the resulting oxicarbide being carbothermally reduced at 1400°C under inert gas, before decomposition at 1800°C (with CO release). It has been claimed that the glassy silica layer of the SA3® fiber presents a remarkable alkali resistance [4]. SiC crystals are 100-300 nm in size [5]; TEM experiments revealed no second phase at the grain boundaries, but turbostratic carbon has been observed at triple points [6].

EXPERIMENTAL PROCEDURE

Samples

SiC-based fiber's nanophases that are in very low concentration (≤1%) or have low Raman cross sections are sometimes hard to observe. In order to trigger the crystalline growth of the nanophases, which have already been observed using X-rays or TEM analysis [25], as-received fibers were first desized at 500°C in air or at 750°C in vacuum (heating rate : 10°C/min; 1 hour dwell) and then heated to 1600°C (SiC growth temperature [7]; 10°C/min under reducing atmosphere and nitrogen flux with a 10 hours dwell). For this we used a furnace with graphite resistor and felt shields with a ~100 mm chamber.

As-received or heat treated fibers, which were analyzed across their diameter, were first coated using nickel electro-deposition (thickness ~ 2mm) and then polished using SiC plates and diamond pastes of 3, 1 and 1/4 µm.

Corrosion

The corrosion procedure is the one described by Colomban et al. [8]. Fibers are placed in thick Pyrex® glass tubes containing molten sodium nitrate salts (T_m=307°C). Because of the low $NaNO_3$ boiling temperature (T_b=380°C), the partial decomposition of NO_3^- in reddish vapours produces strong oxidizing conditions :

$$NaNO_3 \rightarrow Na^+ + NO_3^- \rightarrow Na^+ + NO_2 + \tfrac{1}{2} O_2 (g) + e- \qquad (1)$$

After cooling, the solid nitrate is eliminated by a continuous and slow water flux.

Raman microspectrometry

A LabRAM "Infinity" micro-spectrometer (Jobin-Yvon–Horiba, France) was used in this study. It was equipped with a back-illuminated Spex CCD matrix (2000 x 256 pixels) cooled by Peltier effect (200 K) and two lasers : a He-Ne and a frequency doubled Nd:YAG. It also included Notch filters.

SiC AND CARBON PHASES : RAMAN FINGERPRINTS

Raman lines provide information on the structural parameters of the material [9]. The peak position indicates the chemical nature of the analyzed phase. The area is related to its concentration, and the wavenumber shift gives information on both the nature of the species (specially for resonant lines like D band at ca 1330 cm^{-1}) and the strain level [10]. Finally, the band width is related to the phase short-range order.

Raman peaks attribution

SiC structures alternate layers of Si and C atoms. Two consecutive layers form a bilayer which is named "h" if it is deduced from the one below by a simple translation. If not, when an additional 180° rotation (around the Si-C bond linking the bilayers) is necessary to get the superposition, then the bilayer is named "k" [11]. The "k" only stacking is the reference β-SiC cubic symmetry with main Raman peaks centered at 796 and 972 cm^{-1} corresponding to the transverse (TO) and longitudinal (LO) optic modes, respectively [12]. Any other definite stacking sequence is called α-SiC and has either hexagonal or rhombohedral lattice symmetry.

Raman spectra decomposition

The spatial correlation model describes the crystalline quality by introducing the coherence length (L_0) as a parameter. L_0 is defined as the average extension of the homogeneous domains in the material[13, 14] and is therefore expected to depend on corrosion. According to the model, the intensity $I(\bar{\nu})$ of a Raman peak at the wavenumber $\bar{\nu}$ can be written as :

$$I(\bar{\nu})\ \alpha \int_{q=0}^{q=1} e^{-\frac{k_{BZ}^2 \times (q-q_0)^2 \times L_0^2}{16 \times \pi^2}} \times \frac{dq}{[\bar{\nu}-\bar{\nu}(q)]^2 + \left(\frac{\Gamma_0}{2}\right)^2} \qquad (2)$$

where k_{BZ} is the Brillouin zone limit and q is a reduced wave vector, expressed in units of $2\pi/a$ (a being the lattice constant). Γ_0 is the half width of the Raman peak and the exponential function represents a Gaussian spatial correlation function. $\bar{\nu}$ (q) is the modes dispersion function, which can be deduced from neutron scattering measurements, from calculations or, in the specific case of SiC, inferred from the data on polytypes modes : each one corresponds to 3C-SiC vibration in a specific point along the [111] direction of its reciprocal space[15].

RESULTS AND DISCUSSION

Figs. 1a and 1b show the surface of as-received and 1600°C heat treated NLM® fibers respectively. Comparison is made with $NaNO_3$-corroded Tyranno SA3® fiber in Fig. 1c

As-received NLM® fiber's surface is smooth and homogeneous. Corresponding Raman spectra present large carbon components indicating a low short-range order and very weak SiC signal. However, 1600°C heat treated fiber's surface is covered with ~500 nm crystals, the longest ones exceeding 10 μm. Raman spectra recorded on the latter present very weak carbon signal but strong α and β SiC signals at ca 772, 800 and 969 cm^{-1}. Because of its width and its location between the two optic modes, the broad line at ca 890 cm^{-1} i.e. at a wavenumber close to (800+966)/2=883 cm^{-1} is assigned to amorphous (non-polar) SiC. It is fit with the spatial correlation model, which gives information on the size of the smallest SiC grains. The broad bands at ca 550 and 1080 cm^{-1} could be assigned to a glassy silica signal.

Fig. 1c shows the Raman spectrum of 40min-$NaNO_3$-corroded SA3® fiber. Sharp lines at ca 728, 766 and 828 cm^{-1} (this band clearly appears in other spectra across the fiber) correspond to specific α-SiC polytypes and are then fit with Lorentzian curves. However, the two optic modes, which are centered at ca 795 (TO) and 969 cm^{-1} (LO), are fit according to the spatial correlation model.

A model based on the ratio of the two main carbon Raman peaks (I^D/I^G) has been proposed [16] to calculate the size of short-range ordered vibrational units in carbon moieties. This model takes the relative Raman efficiency d of the D^{1350} and G^{1600} bands into account. Other parameters are R, the ratio of atoms on the surface of each grain with respect to the bulk, e_t the surface thickness and Lg, the coherent length (~ the grain size of Tuinstra and Koenig model [17]). Assuming a spherical shape of all grains (Eq. 3) :

$$d \times R \approx d \times \left[\left(1 - \frac{2 \times e_t}{L_g} \right)^{-3} - 1 \right] \tag{3}$$

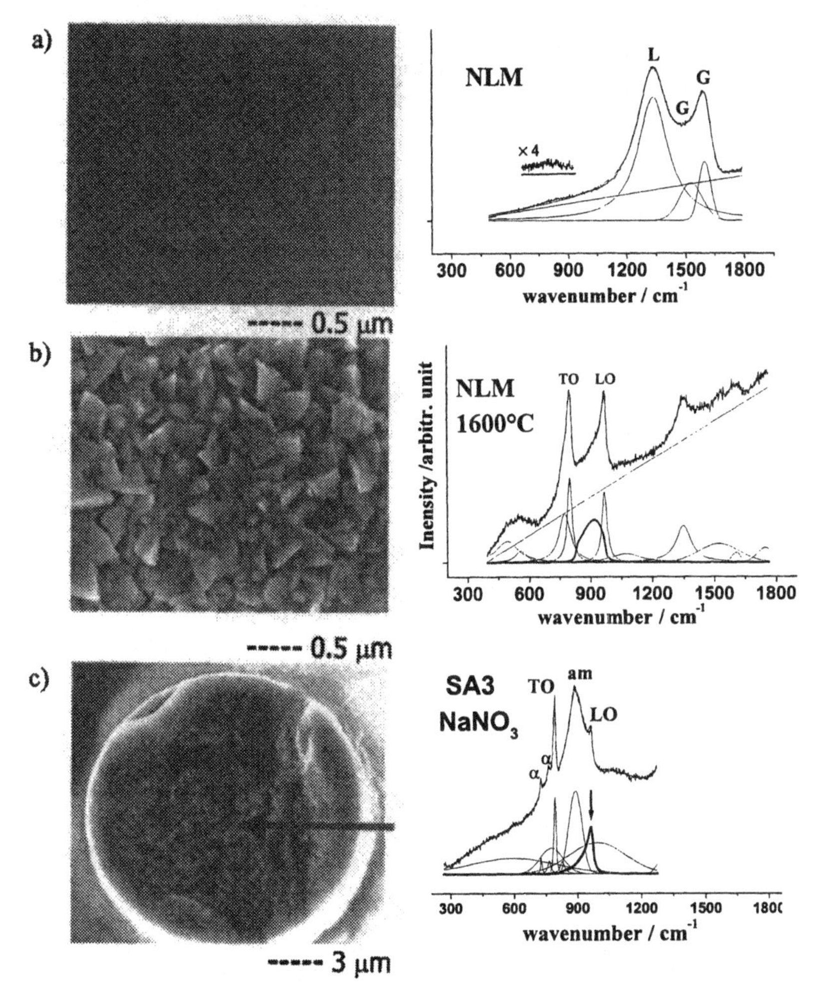

Figure 1: SEM (left) images and Raman spectra (right) of a) as-received NLM® fiber (L: Lorentzian, G: Gaussian); b) 1600°C heat treated NLM® fiber; c) 40min $NaNO_3$-corroded SA3® fiber section (the Raman spectrum was recorded on the fiber's core). The different spectral components are shown (am: amorphous SiC).

Fig. 2 displays the evolution of the ratio I^D/I^G across as-received and 40min $NaNO_3$-corroded NLM® and SA3® fiber's sections. The ratio varies randomly around its average value in NLM® section indicating a homogeneous size distribution of the carbon moieties. The corresponding calculated size is shown on Fig. 3 and is about 0.4 nm. This homogeneity remains after the corrosion.

In contrast, as-received SA3® fiber shows a regular increase of the ratio from the skin to the core revealing smaller carbon moieties in the core. The image of the carbon grains size distribution is shown on Fig. 3. While 1 nm grains are observed on the fiber's core, 2-3 nm grains are observed on the skin. This may be linked to a difference in the crystal growth process during the elaboration. In addition, the action of molten $NaNO_3$ leads to the carbon disappearing in the core area (Fig. 2).

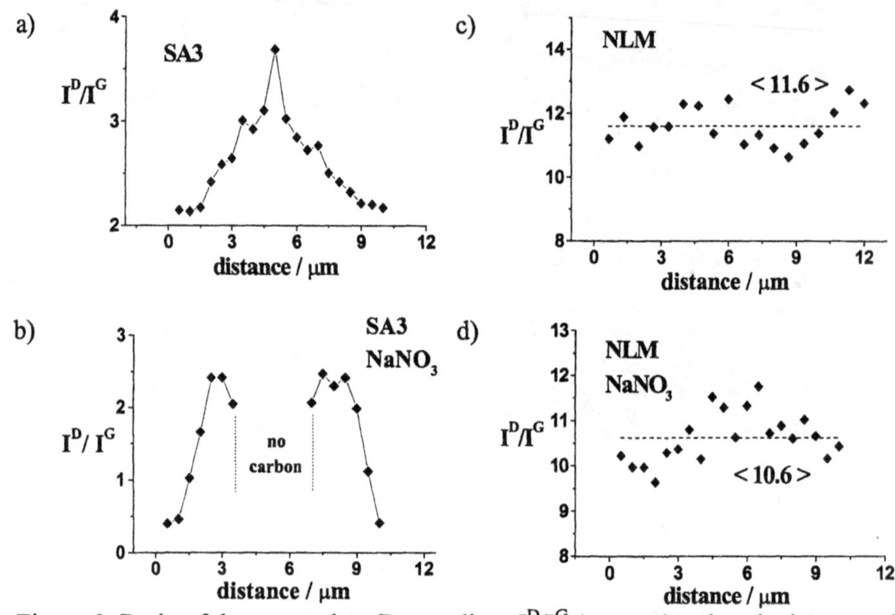

Figure 2. Ratio of the two carbon Raman lines I^D/I^G (proportional to the inverse of the carbon moieties size) across the sections of a) as-received SA3®; b) 40min $NaNO_3$-corroded SA3®; c) as-received NLM®; 40min $NaNO_3$-corroded NLM®.

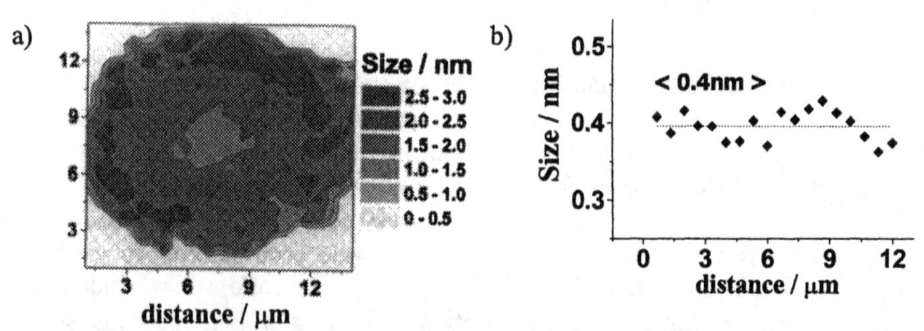

Figure 3. a) Image of the calculated size of the carbon moieties in as-received SA3® fiber; b) Calculated size of the carbon moieties across the as-received NLM® fiber from the I^D/I^G model [16] (see Fig. 2c).

CONCLUSION

Raman microspectrometry was used to analyze whole fibers sections. NLM-Nicalon® fibers were found to be very homogeneous across the fiber regarding the carbon moieties size and reactivity. Indeed, the molten $NaNO_3$ action on the fiber's section revealed no core/skin difference. On the contrary, as-received Tyranno SA3® fiber showed nanostructure and carbon concentration differences across the fiber. The spatial correlation model showed that SiC nanophases are much smaller and less crystallized in SA3® fiber's core than in its skin, indicating a difference in the crystal growth process during the elaboration. Furthermore, non-graphitic carbon moieties which concentrate in the fiber's core are much more sensitive to $NaNO_3$ corrosion. This is attributed to the nature of these carbon moieties and to their location (at triple points and grain boundaries), which constitutes a direct diffusion pathway. Then, Raman microspectrometry appears to be a very efficient analysis method to probe heterogeneous and low crystalline systems, especially in the resonant conditions.

ACKNOWLEDGEMENTS

The authors wish to thank De. Baron for developing the SCM-based peak fitting software. Dr G. Gouadec is acknowledged for many fruitful discussions.

REFERENCES

[1]Ph. Colomban, "Raman Study of Nanophases", *Proc. of the 27th Annual Cocoa Beach Conference & Exposition on Advanced Ceramics & Composites January 26-31, Ceram. Eng. & Sci. Proc* (2003).

[2]R. Bodet and J. Lamon, "Comportement en Fluage de Fibres Céramiques à Base de SiC", *Silicates Industriels* **1-2** (1996).

[3]L. Porte and A. Sartre, "Evidence for a Silicon Oxycarbide Phase in the Nicalon Silicon Carbide Fibre", *J. Mater. Sci.* **24** 271 (1989).

[4]Ishikawa, Y. Kohtoku, K. Kumagawa, T. Yamamura and T. Nagasawa, "High-strength alkali-resistant sintered SiC fibre stable to 2200°C", *Nature* **391-6669** 773-75T (1998).

[5]K. Kumagawa, H. Yamaoka, M. Shibuya and T. Yamamura, "Fabrication and Mechanical Properties of New Improved Si-C-(O) Tyranno Fiber", *Ceram. Eng. & Sci. Proc* **19** [3] 65 (1998).

[6]M. H. Berger, N. Hochet and A. R. Bunsell, "Microstructure and High Temperature Mechanical Behavior of New Polymer Derived SiC Based Fibers", *Ceram. Eng. & Sci. Proc* **19** [3] 39 (1998).

[7]S. Karlin and Ph. Colomban, "Micro-Raman study of SiC fibre-oxide matrix reaction", *Composites Part B* **29B** 41-50 (1998).

[8]Ph. Colomban, G. Gouadec and L. Mazerolles, "Alkaline corrosion of SiC and Carbon Fibers surface. A Raman and electron microscopy study." *Proc. of the 103th Annual American Ceramic Society Conference and Exposition may1-3, Ceramics Transaction* **128** (2001).

[9]I. Kosacki, V. Petrovsky, H. U. Anderson and Ph. Colomban, "Raman Spectroscopy of Nanocrystalline Ceria and Zirconia Thin Films", *J. Am. Ceram. Soc.* **85** [11] 2646-50 (2002).

[10]Ph. Colomban, G. Gouadec and L. Mazerolles, "Raman analysis of materials corrosion : the example of SiC fibers", *Materials and Corrosion* **53** 306-315 (2002).

[11]W. J. Choyke and G. Pensl, "Physical properties of SiC", *MRS Bull.* **22** [3] 25-29 (1997).

[12]H. Okumura, E. Sakuma, J. H. Lee, H. Mulkaida, S. Misawa, K. Endo and S. Yoshida, "Raman Scattering of SiC : Application to the Identification of Heteroepitaxy of SiC Polytypes", *J. Appl.Phys.* **61** [3] 1134-1136 (1987).

[13]P. Parayanthal and F. H. Pollak, "Raman Scattering in Alloy Semiconductors: "Spatial Correlation Model"", *Phys. Rev. Lett.* **52** [20] 1822-25 (1984).

[14]S. Rhomfeld, M. Hundhausen and L. Ley, "Raman Scattering in Polycrystalline 3C-SiC : Influence of Stacking Faults", *Phys. Rev. B* **58** [15] 9858-62 (1998).

[15]D. W. Feldman, J. H. Parker, W. J. Choyke and L. Patrick, "Phonon Dispersion Curves by Raman Scattering in SiC, Polytypes 3C, 4H, 6H, 15R, 21R", *Phys. Rev.* **173** [3] (1968).

[16]Ph. Colomban and G. Gouadec, "Non-destructive Mechanical Characterisation of (nano-sized) Ceramic Fibers", *Key Engineering Materials* **206-213** 677 (2002).

[17]F. Tuinstra and J. L. Koenig, "Characterization of Graphite Fiber Surfaces with Raman Spectroscopy", *Comp. Mater.* **4** 492-99 (1970).

STRUCTURE OF NANOCRYSTALLINE BN AND BN/C COATINGS ON SIC

Linlin Chen, Haihui Ye and Yury Gogotsi
Dept. of Materials Science and Engineering
Drexel University
Philadelphia, Pennsylvania 19104

Michael J. McNallan
Dept. of Civil & Materials Engineering
University of Illinois at Chicago
Chicago, Illinois 60607

ABSTRACT

BN coatings have been synthesized on SiC powders and SiC fibers by a novel carbothermal method. The coatings are produced below 1200°C at atmospheric pressure, they are uniform, and they neither bridge the fibers nor change the surface quality of the fibers. BN and BN/C films were produced. High-resolution transmission electron microscopy (HRTEM), electron energy loss spectroscopy (EELS), scanning electron microscopy (SEM) and Raman microspectroscopy were employed for the microstructural study of these coatings. Amorphous BN and various BN polytypes were found in the coatings by TEM. This process can be used to produce interfacial coatings on SiC fibers for ceramic matrix composites.

INTRODUCTION

Fiber reinforced ceramic matrix composites (CMCs) and continuous fiber ceramic composites (CFCCs) have a wide range of potential applications in heating equipments and chemical industries[1] for their increased toughness over their monolithic counterparts. However, much of the advantage gained by the presence of fibers is often offset by the degradation of mechanical properties at elevated temperatures in oxidizing environments[2]. Therefore, it is important to introduce a protective layer on the fibers as a barrier to block the deleterious interphase reaction between the fibers and the matrix. Also the interface coating can serve as a compliant layer between the fiber and matrix to facilitate load transfer and improve the toughness via fiber pull-out and crack deflection to avoid catastrophic fibrous fracture failure. In many of these applications, structurally and chemically well-defined films are required[3].

Carbon and BN are the most commonly used coatings for CMCs, especially for SiC/SiC composites[4-7]. However, carbon interface coatings can be easily oxidized in air at high temperatures[8-9]. Boron nitride has received considerable attention due to its good mechanical, electrical, optical and chemical properties over a wide range of temperatures[10-11]. Moreover, when BN is oxidized to form B_2O_3, it can in turn react with SiO_2, which is produced by oxidation of SiC fiber core or SiC matrix, to form a glassy protective layer at the interface of the CMCs. The BN coating of 0.2-0.5μm can serve as a weak fiber/matrix interface to promote the fiber debonding and pullout[1]

Boron nitride (BN), like carbon, has cubic (*c*-BN), wurzite (*w*-BN), hexagonal (*h*-BN) and rhombohedral (*r*-BN) modifications, which correspond to diamond (zinc-blende form), hexagonal diamond (wurzite form), and hexagonal and rhombohedral graphite, respectively[12-14]. *c*-BN and *w*-BN are hard, dense phases and are bonded through strong sp^3 hybridized σ bonds, and they both consist of tetrahedrally coordinated boron and nitrogen atoms. *c*-BN is arranged in a three-layer (ABCABC...) and *w*-BN arranged in a two-layer (ABAB...) stacking sequence[15]. *h*-BN and *r*-BN are softer, graphite-like phases, and both consist of hexagon layers that arrange in two-(ABAB...) and three-layer (ABCABC...) stacking sequences, respectively. For *h*-BN and *r*-BN, the in-plane atoms are both bonded through localized sp^2 hybridization, while the out-of-plane layers are bonded by delocalized orbitals with weak Van der Waals forces[15]. Structurally, direct compression along the c-axes of *h*-BN or *r*-BN yields *w*-BN and *c*-BN (the so-called puckering mechanism), respectively[16].

Nowadays, BN thin coatings on ceramic fibers for CMCs are primarily produced by chemical vapor deposition (CVD)[17-18]. However, the cost and the difficulties in controlling the CVD process limit the potential applications and it is also difficult to avoid fiber bridging in CVD deposition. A novel carbothermal synthesis of BN coatings below 1200°C in atmospheric pressure had been proposed and developed by the authors[19]. It offers a simple and cost-effective way to control the thickness of the BN coatings. The purpose of this study is to provide a chemical and structural characterization of BN coatings synthesized on SiC materials by this method.

EXPERIMENTAL PROCEDURE

Materials

β-SiC powder, supplied by Superior Graphite Co., USA, with purity of 99.8%, 1μm particle size, and Tyranno ZMI SiC fibers (56% silicon, 34% carbon, 9% oxygen and 1% zirconium), produced by UBE Industries of Japan, were used as the raw materials for the synthesis of BN coatings by the proposed method[19]. β-SiC powder was used as a model system to understand the process mechanism due to its easier sample preparation for XRD and TEM studies. Tyranno fibers were then used to determine the feasibility of this method in composite systems.

BN Coating Synthesis

The synthesis process includes three main steps: chlorination, infiltration and nitridation. The detailed process for each step is presented with respect to the different types of SiC.

(1)*Chlorination and Infiltration:* (i) Powders—β-SiC powder was treated in pure chlorine at a flow rate of 10 standard cubic centimeters per minute (sccm) at 1000°C for 3 hours in a quartz tube furnace with diameter of 2.5 cm. The carbide derived carbon (CDC) powders obtained by the chlorination were mixed together with H_3BO_3 powders with a C/B stoichiometric ratio of 9:4. Then the mixture was ground in a mortar to achieve a grain size ~1μm. Infiltration of the powders by H_3BO_3 occurred during the heating for nitridation because B_2O_3 melts at 450°C. (ii) Fibers—Tyranno ZMI SiC fibers (~10μm in diameter) were treated in pure Cl_2 with gas flow of 10sccm for 3 hours at 550°C and 650°C at atmospheric pressure to form carbon coatings on fibers with the thickness of ~ 0.15μm and ~1.5μm, respectively. The CDC coated fibers were placed in the vacuum infiltration chamber and pumped down for about 30 minutes. Then they were infiltrated with a saturated H_3BO_3 solution at 100°C. Cold distilled water was used to wash out the excess of H_3BO_3 from the fiber surface at room temperature. To achieve good infiltration result in nanoporous carbon coating, this step was repeated two or three times

(2)Nitridation: Infiltrated CDC-coated SiC samples (powders and fibers) were loaded in a quartz boat and put into a horizontal quartz tube furnace with the inner diameter of 2.5 cm. Before each experimental run, the furnace was purged with argon for at least 30 minutes. Then the furnace was heated to 1150°C at a rate of 10°C/min with ammonia (grade 4: purity 99.99%, BOC gases) flowing into the reaction tube at a flow rate of 10sccm. The sample was held at this temperature for 60-80 minutes to ensure the completion of the reaction, and then cooled down in the furnace under the ammonia flow for protection from hydration.

TEM Sample Preparation

BN-coated CDC SiC powders were dispersed in methyl or isopropyl alcohol, assisted by ultrasonication, and deposited on Lacey carbon film TEM grids.

The BN-coated Tyranno ZMI SiC fibers were prepared for TEM analysis using the following route. (i) BN-coated fiber bundles were horizontally imbedded in liquid epoxy resin with a thickness less than 5mm in a 3cm inner diameter plastic mould. (ii) After overnight drying, the margin of such resin disc was cut with a diamond wire saw. A 3mm-diameter disc with embedded fibers can be made with less than 80 μm thickness by polishing with SiC sand paper (#1200 grade). (iii) Gatan dimple grinder (Model 656) was used to dimple the ground side of the disc to a thickness of ≤30μm. (iv) The specimen was finally thinned by an ion-mill (Precision Ion Polishing System, Model 691, PIPS™ V 4.32) until appearance of a hole in the center. At this point, the edges of the hole as well as the embedded fibers are transparent for electrons (thickness: 10–100 nm). The ion-mill was operated at 4 kV and 1 mA, with the incident angle of the argon-ions of 8° in the single sector thinning mode.

Characterization

The composition and structures of the samples nitrided under various conditions were analysized by environmental scanning electron microscopy (ESEM: FEI XL-30), energy dispersive spectroscopy (EDS), high-resolution transmission electron microscopy (HRTEM; JEOL 2010F, operated at 200KV) and elcctron energy loss spectroscopy (EELS), respectively.

RESULTS AND DISCUSSION

BN Coatings on CDC *β*-SiC Powders

Pure β-SiC powders were completely converted into carbon by chlorination in pure Cl_2 at 1000°C for 3 hours. BN coatings were obtained after the nitridation of H_3BO_3-infiltrated CDC powders at 1165°C for 60 minutes in ammonia.

The chemical composition of BN coatings was characterized by EDX and EELS. Three distinct layers were detected by the EELS element mapping as shown in Figure 1. The outermost layer of the powder is pure BN coating with an average thickness of 50-70nm, in which the carbon was totally consumed during the reaction. The intermediate layer is a mixture of BN and carbon with a thickness of 75-110nm and the inner layer (core) of the particle is unreacted carbon. The central part of the particle could not be fully represented by the EELS mapping due to the thickness limitation (Fig. 1c). The total thickness of BN coating is around 120-180nm. Formation of BN and mixed BN/C layers most probably is attributed to the maximization of energetically favorable C-C and B-N bonds, rather than the B-C and N-C bonds[20-22].

Figure 2 shows the HRTEM images of such BN-coated CDC powders. The BN layer was mainly composed of *h*-BN and amorphous BN. The mechanism of formation of BN could be understood by HRTEM and EELS analysis. The inner layer of the coating at BN/C interface is mainly composed of amorphous BN. A certain amount of *h*-BN crystals, with spacing d_{002}=0.336nm as shown in Fig. 2b, appeared in the middle of the BN coatings and crystalline *h*-BN became dominant in the surface layer of the coatings. A very small amount of cubic BN crystals with spacing d_{111}=0.209nm was also detected at the interface with the CDC layer (Fig. 2c). The mechanism of its formation is probably attributed to the nanocrystalline diamond growth upon chlorination of SiC[23-24]. The nanodiamond may help to induce the formation of cubic BN crystals acting as a template or seed. The presence of cubic BN nanocrystals directly adjacent to carbon in Fig. 2c supports this hypothesis. It can be assumed that the amorphous and diamond-structured BN formed by the reaction with ammonia at the C/BN interface, and slowly transformed to the more stable hexagonal modification as the reaction front propagated toward the particle core during nitridation. The presence of hexagonal BN in the intermediate layer of BN coatings suggests this mechanism. The increased temperature during nitridation makes the conversion from *a*-BN to *h*-BN more kinetically favorable[25]. This is also consistent with the reported BN growth sequence: *a*-BN→*h*-BN[26-27]. Small amounts of boron oxides generated from the reactants in the beginning of the nitridation also help to form the hexagonal-structured boron nitride. The same phenomenon has been reported during the synthesis of BN by CVD[28].

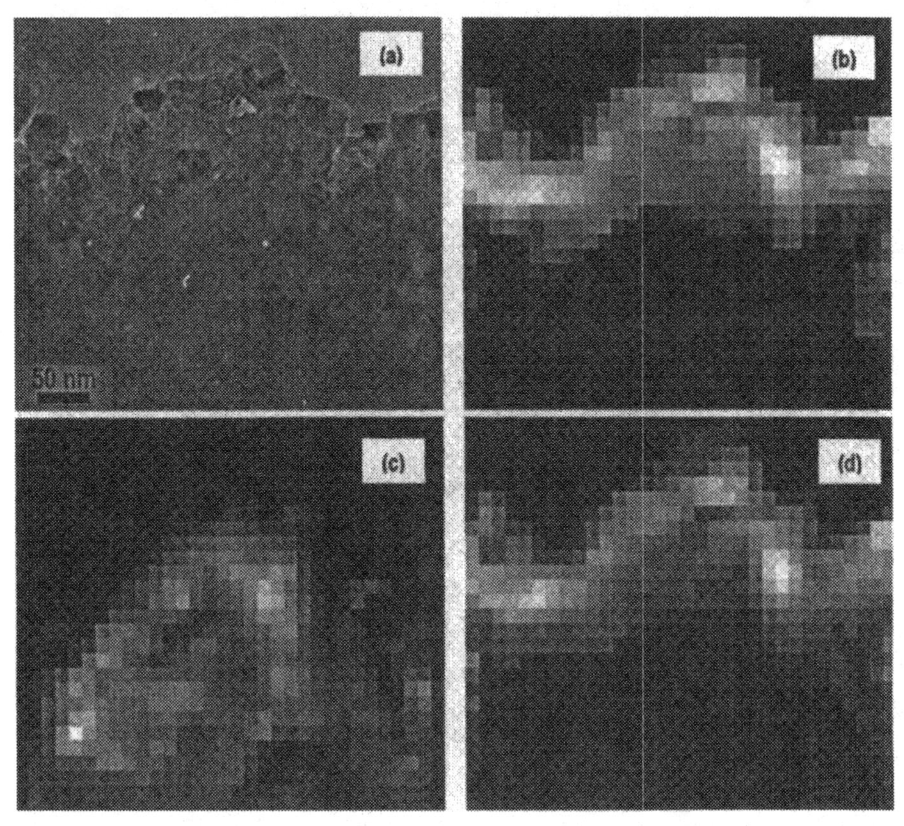

Figure 1: EELS elemental mapping of the BN-coated CDC β-SiC powders by nitridation in NH_3 at 1165°C for 60 minutes. (a) TEM image; (b) Boron map; (c) Carbon map; (d) Nitrogen map.

BN Coatings on CDC-coated Tyranno ZMI SiC Fibers

The fibers used for preparation of the TEM samples had a 200nm-1.5μm thickness CDC layers and were nitrided at 1150°C in ammonia for 80 minutes.

The introduction of CDC layer does not only help to facilitate the reaction of NH_3 with B_2O_3, but also protects SiC fibers from degradation due to the oxidation by water vapor generated in the nitridation reaction. CDC layer is also used to control the thickness of BN coatings. BN-coated SiC fibers with a thin CDC intermediate layer are smooth, uniform and have good mechanical properties[29]. No fiber bridging has been observed in the nitrided fiber bundles and fabrics as shown in Figure 3.

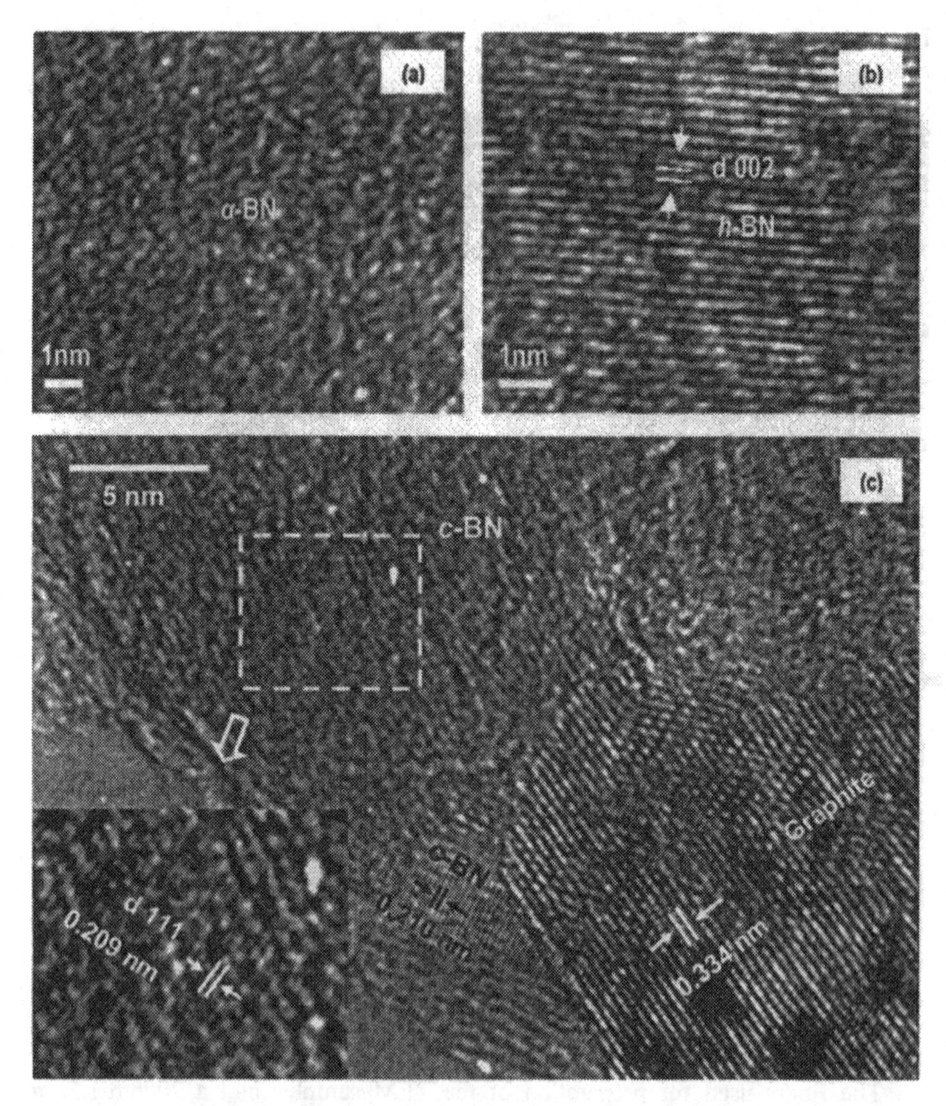

Figure 2: TEM images of BN crystals formed by the nitridation of CDC β-SiC powder at 1150°C for 80 minutes. (a) Amorphous BN; (b) Hexagonal BN; (c) Cubic BN.

A typical homogenous BN coating with good adherence to the fiber core is clearly seen in Figure 4a. The coating formed on such CDC-coated SiC fibers consists of the BN layer on the top and intermediate layer composed of the BN together with some unreacted carbon. TEM image of the crystallites formed in the coating layers are shown in Figure 4b. It can be seen that part of the surface coating was lost during the ion-milling. Amorphous carbon, amorphous BN and some BN crystals were found in the coating. Details of their crystal structures can be seen in the HRTEM image in Figure 4c. Amorphous BN and hexagonal BN are still the main components in the coatings. Small amount of cubic BN crystallines were also found at the interface of BN/CDC, which can be clearly seen in the Fig.4c. The formation of the nanocrystalline cubic BN was most probably induced by the nanocrystalline diamond-structured carbon in the CDC layers, which were formed by the chlorination[24] and contributed to the increase of the Young's modulus from 190GPa of the as-received fibers to 245GPa of the thin carbon-coated fibers[29]. Similar to carbon, at very small grain size (less than 5-10nm), cubic BN may be energetically favored over *h*-BN[30]. This is also consistent with the aforementioned results of the BN coatings on SiC powder.

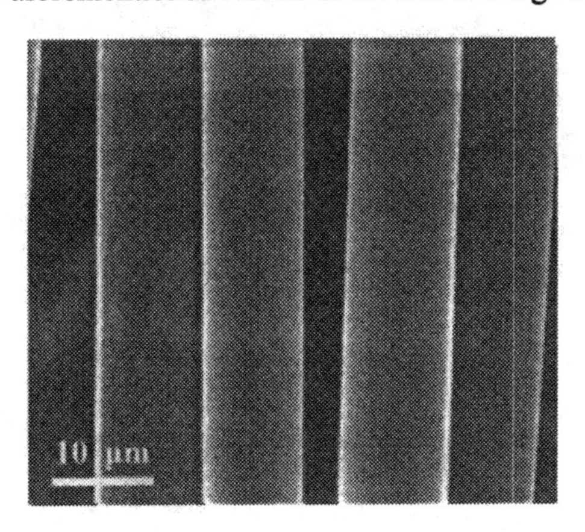

Figure 3: SEM image of the BN-coated Tyranno ZMI SiC fiber bundles. (0.15μm CDC layer, nitridation at 1150°C for 60 minutes)

It is important to note that Tyranno ZMI fiber contains some amorphous silica (~5%) and it was not removed by chlorination or nitridation. It was not crystallized and stayed in amorphous phase. Presence of SiO_2 may improve oxidation resistance of BN as shown in previous studies[31], and provide better oxidation protection to the fibers, and CMCs produced using these fibers.

This proposed method allows synthesis of the thermally and mechanically stable BN coatings on a variety of carbide materials (Fig. 5). They can be used as insulating, protecting and tribological coatings for a variety of applications.

CONCLUSIONS

1. HRTEM and EELS analysis show that the BN coatings synthesized by the nitridation of H_3BO_3-infiltrated CDC coatings on SiC powders and fibers are smooth and uniform. The coatings adhere well to the fiber core with no bridging in fiber bundles.

2. Amorphous BN and hexagonal graphitic BN nanocrystals have been detected in the coatings with a gradient of these phases from the BN/C interface to

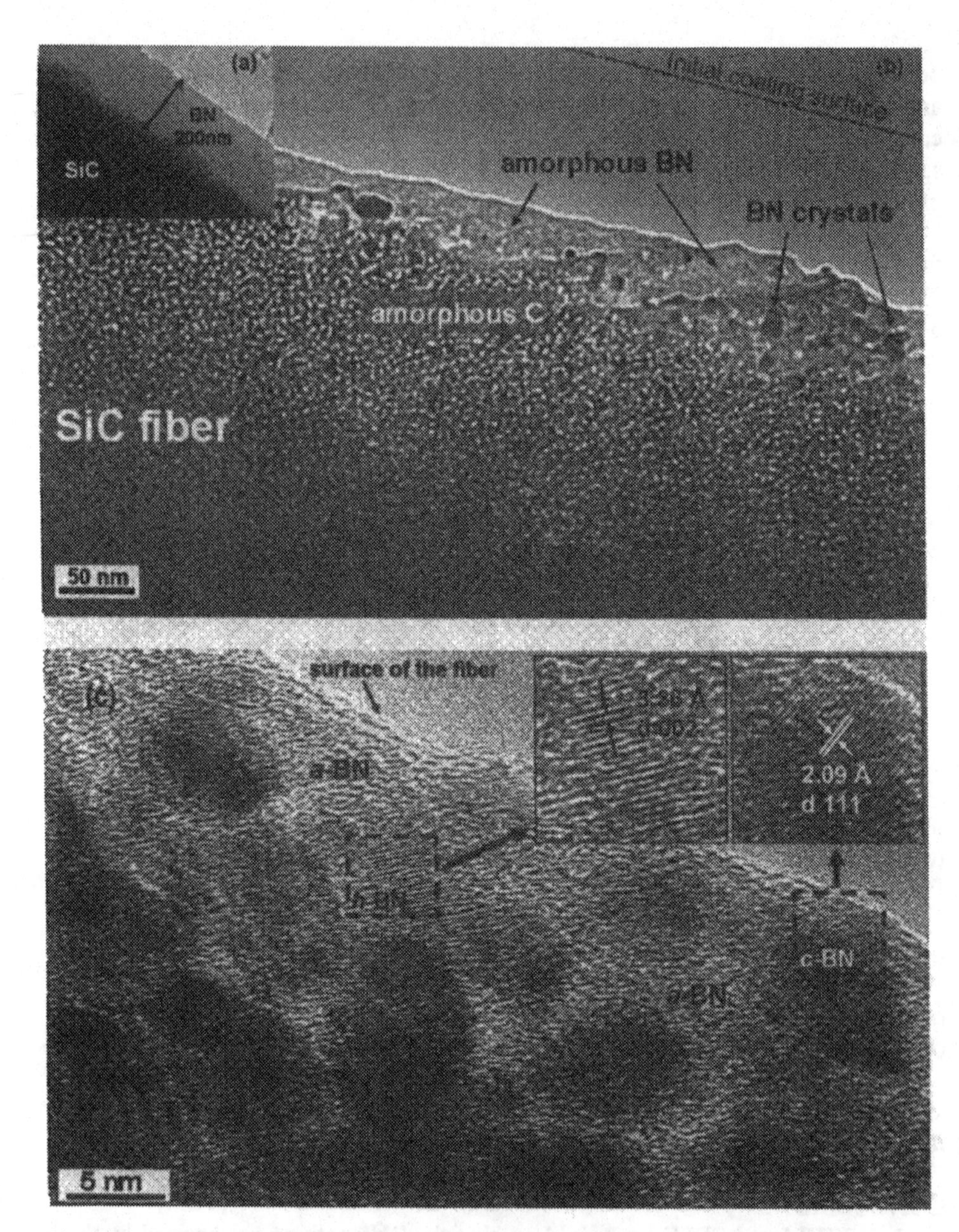

Figure 4: TEM images of the BN-coated Tyranno ZMI SiC fiber. (200nm thick CDC layer, nitridation at 1150°C for 80 minutes). (a) TEM image of the BN-coated fiber shows the uniform BN coatings with good adherence to the fiber core; (b) TEM image of the nanocrystals embedded in amorphous phase in the coating layers on the Tyranno fibers; (c) HRTEM image shows the structures of BN nanocrystals in the coatings.

the surface. Also small amounts of the cubic BN nanocrystals were found at the BN/CDC interface.

3. The BN coating thickness can be controlled with nanometer accuracy in the range from several nanometers to several micrometers by controlling the CDC layer thickness as described in reference[23].

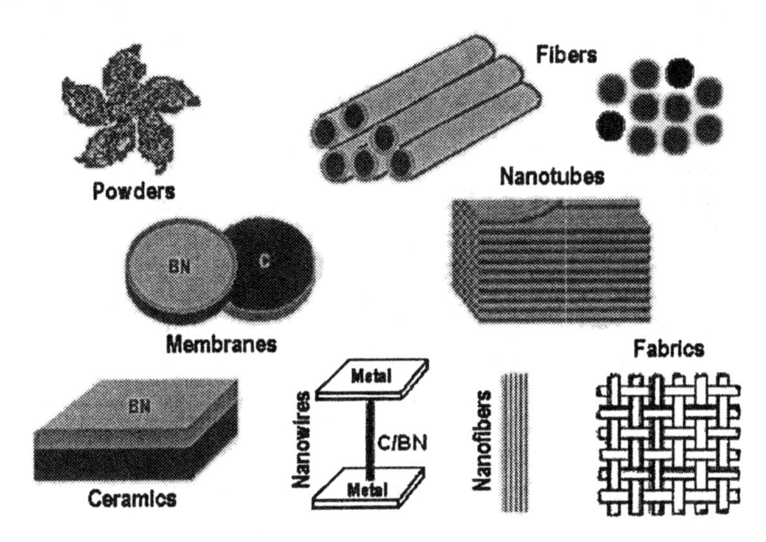

Figure 5: Schematic of the potential applications of BN coatings on carbide.

ACKNOWLEDGEMENTS

We gratefully acknowledge Dr. Alexei Nikitin and Ms. Beth Carroll for experimental assistance and useful discussions and Mr. David Von Rohr (all in Dept. of Materials Science and Engineering, Drexel University) for his help in the SEM characterization. Thanks also to Dr. J. Schwarz of SSG Precision Optronics Corporation for the supply of SiC fibers and Dr. I. Barsukov of Superior Graphite Corporation for providing SiC powders. This work was supported by NASA via a SBIR grant to SSG Precision Optronics Corporation.

REFERENCES

[1]C.X. Campbell and M.G. Jenkins, "In-situ Determination of Constituent Properties and Performance in an Oxide-oxide CFCC," *Ceram. Eng. Sci. Proc.*, **20**[3] 545-54 (1999).

[2]P.F. Tortorelli, C.A. Wijayawardhana, L. Riester and R.A. Lowden, "Oxidation Effect on Nextel-reinforced SiC," *Ceram. Eng. Sci. Proc.*, **15**[4] 262-71 (1994).

[3]T. Klotzbücher, W. Pfleging, M. Mertin, D.A. Wesner and E.W. Kreutz, "Structure and Chemical Composition of BN Thin Films Grown by Pulsed-laser Deposition (PLD)," *Applied Surface Science*, **86**, 165-69 (1995).

[4]R.N. Singh, and M.K. Brun, "Effect of Boron Nitride Coating on Fiber-Matrix Interactions," *Ceram. Eng. Sci. Proc.*, **8** [7-8] 636-643 (1987).

[5]M.A. Kmeltz, J.M. Laliberte, W.S. Willlis, S.L. Suib and F.S. Galasso, "Synthesis, Characterization, and Tensile Strength of CVI SiC/BN/SiC Composites," *Ceram. Eng. Sci. Proc.*, **12** [9-10] 2161-74 (1991).

[6]R. Nasain, O. Dugne, A. Guette, J. Sevely, C.R. Brosse, J.P. Rocher and J. Cotteret, "Boron Nitride Interphase in Ceramic-Matrix Composites," *J. Am. Ceram. Soc.*, **74** [10] 2482-88 (1991).

[7]R.D. Veltri and F.S. Galasso, "Chemical-Vapor-Infiltrated Silicon Nitride, Boron Nitride, and Silicon Carbide Matrix Composites," *J. Am. Ceram. Soc.*, **73** [7] 2137-40 (1990).

[8]C.F. Windisch Jr., C.H. Henager Jr., G.D. Springer and R.H. Jones, "Oxidation of the Carbon Interface in Nicalon-Fiber-Reinforced Silicon Carbide Composite," *J. Am. Ceram. Soc.*, **80** [3] 569-74 (1997).

[9]Y.G. Gogotsi, and M. Yoshimura, "Low-Temperature Oxidation, Hydrothermal Corrosion, and Their Effects on Properties of SiC (Tyranno) Fibers," *J. Am. Ceram. Soc.*, **78** [6] 1439-50 (1995).

[10]S.P.S. Arya, and A.D. Amico, "Preparation, Properties and Applications of Boron Nitride Thin Film," *Thin Solid Films*, **157**, 267-82 (1988).

[11]P.B. Mirkarimi, K.F. McCarty and D.L. Medlin, "Review of Advances in Cubic Boron Nitride Film Synthesis," *Mater. Sci. Eng*., R, **21**, 47-100 (1997).

[12]S. Ulrich, T. Theel, J. Schwan, and H. Ehrhardt, "Magnetron-sputtered Superhard Materials," *Surf. Coat. Technol*., **97**, 45-59 (1997).

[13]M.B. Mekki, M.A. Djouadi, E. Guiot, V. Mortet, J. Pascallon, V. Stambouli, D. Bouchier, N. Mestres and G. Nouet, "Structure Investigation of BN Films Grown by Ion-Beam-Assisted Deposition by Means of Polarised IR and Raman Spectroscopy," *Surf. Coat. Technol.*, **116-119**, 93-99 (1999).

[14]B.T. Kelly, *Physics of Graphite*, Applied Science, London, U.K., 1981.

[15]P.B. Mirkarimi, K.F. McCarthy and D.L. Medlin, "Review of Advances in Cubic Boron Nitride Film Synthesis," *Mater. Sci. Eng.*, R, **21**, 47-100 (1997).

[16]A.V. Kurdyumov, V.L. Solozhenko and W.B. Zelyavski, "Lattice Parameters of Boron Nitride Polymorphous Modifications as a Function of Crystal-Structure Perfection," *J. Appl. Crystallogr*., **28**, 540-45 (1995).

[17]S.P.S. Arya and A. D'Amico, "Preparation, Properties and Applications of Boron Nitride Thin Films," *Thin. Solid Films,* **157**, 267-82 (1988).

[18]D.P. Stinton, T.M. Besmann and R.A. Lowden, "Advanced Ceramics by Chemical Vapor Deposition Techniques," *Ceram. Bull.,* **67** [2] 350-55 (1988).

[19]L. Chen, H. Ye, Y. Gogotsi and M. J. McNallan, "Carbothermal Synthesis of BN Coatings on SiC," *J. Am. Ceram. Soc.*, in press (2003).

[20]H. Sachdev, R. Haubner, H. Nöth and B. Lux, "Investigation of the c-BN/h-BN Phase Transformation at Normal Pressure," *Diamond Relat. Mater*., **6**, 286-92 (1997).

[21]R. Ma, Y. Bando and T. Sato, "Coaxial Nanocables: Fe Nanowires Encapsulated in BN Nanotubes with Intermediate C Layers," *Chem. Phys. Lett.*, **350**, 1-5 (2001).

[22]H. Nozaki and S. Itoh, "Structural Stability of BC_2N," *J. Phys. Chem. Solids*, **57** [1] 41-49 (1996).

[23]Y. Gogotsi, S. Welz, J. Daghfal and M.J. McNallan, "Formation of Carbon Coatings on SiC Fibers by Selective Etching in Halogens and Supercritical Water," *Ceram. Eng. Sci. Proc.*, **19** [3] 87-94 (1998).

[24]Y. Gogotsi, S. Welz, D.A. Ersoy and M.J. McNallan, "Conversion of Silicon Carbide to Crystalline Diamond-structured Carbon at Ambient Pressure," *Nature,* **411**, 283-87 (2001).

[25]T. Klotzbücher, W. Pfleging, M. Mertin, D.A. Wesner and E.W. Kreutz, "Structure and Chemical Composition of BN Thin Films Grown by Pulsed-laser Deposition (PLD)," *Applied Surface Science*, **86**, 165-69 (1995).

[26]D.J. Kester, K.S. Ailey, R.F. Davis and K.L. More, "Phase Evolution in Boron Nitride Thin Film," *J. Mater. Res.*, **8** [6] 1213-16 (1993).

[27]H.C. Hofsäss, C. Ronning, U. Griesmaier, M. Gross, S. Reinke, M. Kuhr, J. Zweck and R. Fischer, "Characterization of Cubic Boron Nitride Films Grown by Mass Separated Ion Beam Deposition," *Nucl. Instrum. Meth. B,* **106** [1-4] 153-58 (1995).

[28]Y. Matsui, Y. Sekikawa, T. Sato, T. Ishii, S. Isakawa and K. Shii, "Formations of Rhombohedral Boron Nitride, as Revealed by TEM- Electron Energy Loss Spectroscopy," *J. Mater. Sci.*, **16**, 1114-16 (1981).

[29]L. Chen, G. Behlau, M.J. McNallan and Y. Gogotsi, "Carbide Derived Carbon (CDC) Coatings for Tyranno ZMI SiC Fibers," *Ceram. Eng. Sci. Proc.*, in press, (2003).

[30]O.A. Shenderova, V.V. Zhirnov and D.W. Brenner, "Carbon Nanostructures," *Critical Review in Solid State and Mater. Sci.*, **16** [3/4] 227-356 (2002).

[31]Y. Gogotsi and V.A. Lavrenko, "Corrosion of High-Performance Ceramics", pp.190, Springer Press, Berlin, 1992.

PREPARATION OF IRON OXIDE AND IRON OXIDE/SILICON OXIDE NANOPARTICLES VIA WATER-IN-OIL MICROEMULSION

Xujin Bao, Min Lin, Hwee Zhen Koh and Qiuyu Zhang
Institute of Polymer Technology and Materials Engineering (IPTME)
Loughborough University
Loughborough, Leicestershire LE11 3TU
UK

ABSTRACT

A new water-in-oil (W/O) reverse microemulsion system was developed to prepare silica-coated iron oxide nanoparticles, where Igepal CO-520 was used as the surfactant and cyclohexane as the oil phase. As a comparison, coprecipitation method was also employed to synthesize the nanoparicles with hydrothermal post-treatment to enhance the crystallinity of the iron oxide nanoparticles. The iron oxide nanoparticles prepared via the microemulsion and co-precipitation at ambient temperature are magnetite (Fe_3O_4) and have poor crystallinity with the size of 5-10nm and 10-20nm, respectively. The iron oxide particles via microemulsion grew in size to ~60nm after hydrothermal treatment at 200°C and transformed to hematite (α-Fe_2O_3) at 140°C. Large size distribution (14-180nm) was observed for nanoparticles prepared by coprecipitation after post-hydrothermal treatment at 200°C and maghemite (γ-Fe_2O_3) phase was formed at 140°C - 160°C and transformed into the hematite at 200°C. Silicon oxide coated iron oxide nanocomposites were also synthesized in-situ by ammonia-catalyzed tetraethyl orthosilicate (TEOS) hydrolysis and condensation after forming the iron oxide nanoparticles in the microemulsion system. TEM with EDS showed that the iron oxides were coated by a thin layer of silica.

INTRODUCTION

In recent years, there has been considerable interest in the synthesis of nano-sized magnetic particles such as Fe_2O_3 and Fe_3O_4, due to their potential applications as magnetic inks, ferrofluids, magnetic recording, magnetic storage media, and magnetic cell separation and other medical applications [1-5]. However, pure magnetic particles themselves may not be very useful in practical applications as they tend to form large aggregates and they can undergo rapid biodegradation when they are directly exposed to the biological system [6]. A coating is necessary to prevent such problems [7-10]. Furthermore, nanomagnetic composites with core/shell structures could result in spin bias exchange at the boundary of the superparamagnetic particles and, as a consequence of single domain characteristics, lead to enhanced coercivity and apparent ferromagnetic behaviour [11], These coated particles would also be very useful for efficient biomolecule separation and for magnetically guided biosensor applications [6].

A microemulsion is a thermodynamically stable system composed of at least three components: two immiscible components (usually water and oil) and a surfactant [12], which provides a microheterogeneous medium for the generation of nanoparticles. The formation of particles in such systems is controlled by the reactant distribution in the droplets and by the dynamics of interdroplet exchange. The surfactant stabilized micro-cavities provide a cage-like effect that limits particle nucleation, growth and agglomeration. In this report, a new water-in-oil microemulsion system was developed to prepare iron oxide nanoparticles, silicon oxide nano-spheres and iron oxide/silicon oxide nanoparticles with core-shell structure. Transmission electron microscopy (TEM), X-ray diffraction (XRD) and energy dispersive X-ray analysis were employed to study the morphology, particle size and crystallinity of both the uncoated and silica-coated iron oxide particles.

EXPERIMENTAL DETAILS

Materials

All chemicals were purchased from Sigma-Aldrich and were used as receive. A non-ionic surfactant Igepal CO-520 (polyoxyethylene(5)nonylphenyl ether) was used as the surfactant. Iron chloride (II) tetrahydrate 99%, iron chloride (III) hexahydrate 98%, tetraethylorthosilicate (TEOS), and ammonium hydroxide 35% were reactants for the iron oxides and silica coated iron oxides. The oil phase was cyclohexane. Distilled water was used throughout the experiment.

Synthesis of iron oxide, silicon oxide and iron oxide/silicon oxide nanoparticles

The synthesis of the iron oxide nanoparticles was carried out in an Igepal CO-520/cyclohexane (20/80 wt%) reverse microemulsion. Precalculated amounts of iron chloride aqueous solution (Fe^{++}/Fe^{+++} = 1:2 mol%)were added to the microemulsion system with NH_4OH as the base (pH = 9.0). Nitrogen gas was continuously purged during the mixing. Resultant products were left to react under ambient temperature for 24 hours. Silica nanoparticles were prepared by the hydrolysis of tetraethylorthosilicate (TEOS) (H_2O/TEOS = 40 mol/mol) in the same reverse microemulsion as above. Silica coated iron oxide particles were prepared by two steps: iron oxide nanoparticles was first synthesized as described above, then precalculated TEOS was added into the microemulsion. The mixed solution was left to react at ambient temperature for 72 hours. Iron oxide nanoparticles were also synthesized by the coprecipitation of iron salts with NH_4OH. Same amount of $FeCl_2$, $FeCl_3$ and NH_4OH with same concentrations as in microemulsion were mixed and reacted at room temperature for 24 hours. All the nanoparticles synthesised were separated by centrifugation, washed thoroughly with acetone and distilled water to remove any extraneous species (two times with acetone followed by two times with distilled water and finally with acetone again).

Hydrothermal treatment

The hydrothermal treatment of the nanoparticles synthesised was carried out in an autoclave at temperature between 140 and 200°C. The autoclave is an airtight vessel made of stainless steel with a Teflon inner container, and a stainless steel cap. The starting materials had been washed and centrifuged before undergoing the hydrothermal treatment. 1g of nanoparticles and 20g of water was mixed and sealed in the autoclave. The autoclave was kept in a temperature-controlled oven at the reaction temperature for 5h. Internal pressure depends on the temperature.

Characterisation

A transmission electron microscope, JEOL 2000FX, was used to examine iron oxide, silica and silica-coated iron oxide nanoparticles. Samples were made by placing a drop of the solution on a copper grid. The solvent was allowed to evaporate at room temperature. Particle sizes were estimated from TEM pictures. The silica-coated iron oxide particles were further examined using a FEI Tecnai F20 Field Emission Gun Transmission Electron Microscope equipped with thin window Energy Dispersive X-ray Analysis (EDX). The crystallinity of the powder samples was measured using a Bruker D8 X-ray diffractometer with CuK_{α} radiation λ = 0.15406nm at 40kv and 30mA. Data were collected over the 2θ range 10 - 70° with a step size of 0.02° and a count time of 0.5s. Samples to be characterized using XRD were dried and ground into fine powders before measurement.

RESULTS AND DISCUSSION

Iron oxide particles were synthesized via the microemulsion and coprecipitate methods. The size of the nanoparticles synthesized by microemulsion is in the range of 5-10nm. The particles prepared by coprecipitation were 10-20nm (Fig. 1). The introduction of microemulsion to produce nanoparticles bound the particle size within smaller nano-range and facilitate the uniform particles. While the co-precipitation method does not hold a micro-cage in "moulding" the particles morphology and thus produces particles with a relatively larger particle size and a broader particle size distribution. The X-ray diffraction pattern (XRD) indicated these nanoparticles are magnetite (Fe_3O_4) and have poor crystallinity (Fig. 2).

The investigation of the effect of hydrothermal treatment on the crystallinity and phase changes of the iron oxide nanoparticles was carried out in a small autoclave at temperatures from 140°C to 200°C. It is worthy to note that the growth of iron oxides synthesised via microemulsions was from 10nm to ~60nm with rather narrow size distribution after hydrothermal treatment (Fig. 3a), while irregular size distribution (14-180nm) was observed for the iron oxide nanoparticles prepared via coprecipitation, sizes ranges from (Fig. 3b). XRD (Fig. 4) showed that the iron oxide prepared via microemulsions was transformed from magnetite to hematite (α-Fe_2O_3) at hydrothermal temperature above 140°C. In contrast, a maghemite (γ-Fe_2O_3) phase was observed for the nanoparticles prepared via coprecipitation following by hydrothermal treatment at 140°C, 160°C and transformed into hematite (α-Fe_2O_3) at 200°C (Fig. 5).

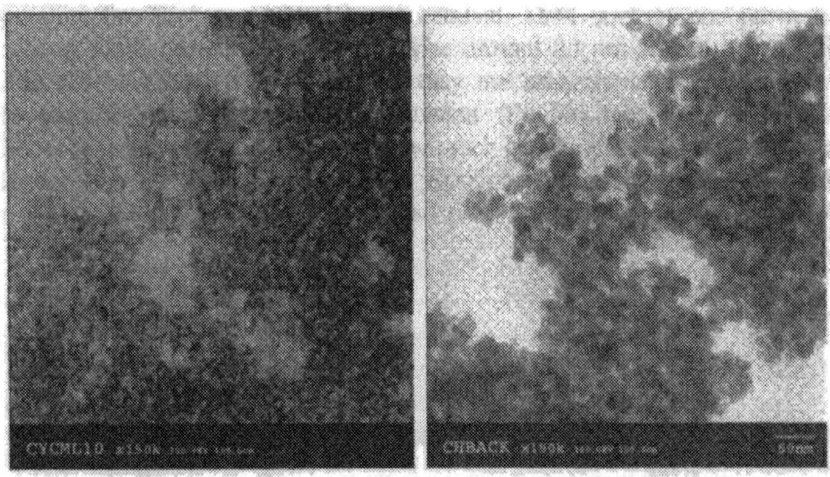

Fig. 1 TEM micrographs of iron oxide nanoparticles prepared by (left) microemulsion and (right) coprecipitated at ambient temperature.

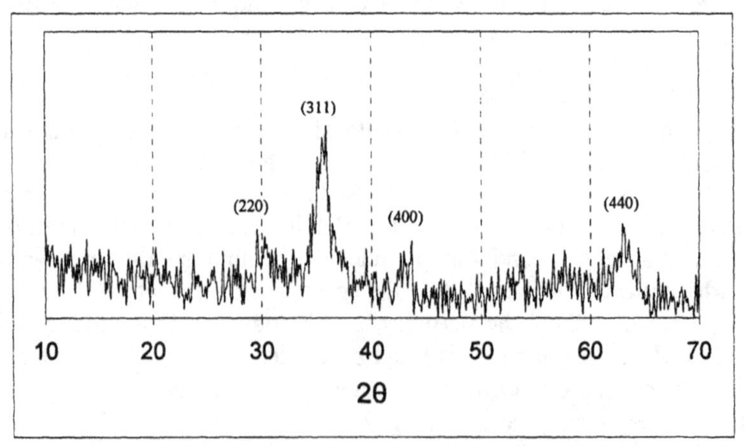

Fig. 2. X-ray diffraction pattern of the iron oxide nanoparticles prepared via microemulsions at ambient temperature.

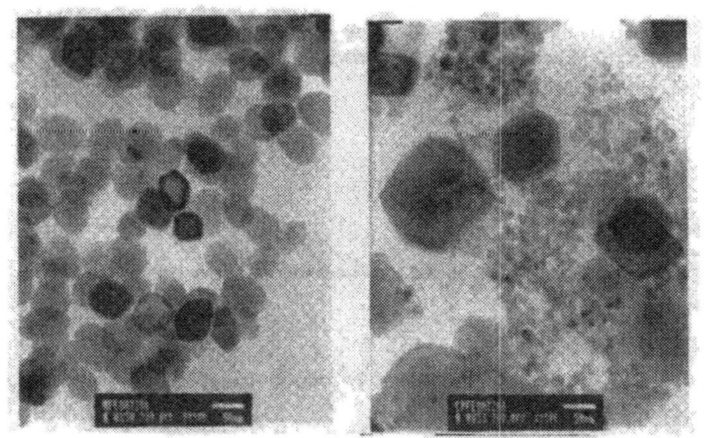

Fig. 3. TEM mirographs of iron oxide nanoparticles prepared by (left) microemulsion and (right) coprecipitation after hydrothermal treatment at 200°C

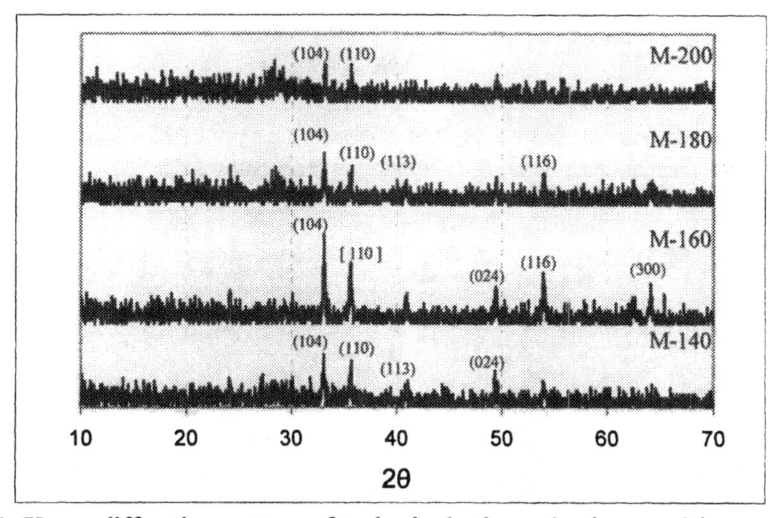

Fig. 4. X-ray diffraction patterns for the hydrothermal microemulsion prepared iron oxide as a function of synthesis temperature as the solution reacted for 5h.

The silicon oxides were synthesized by the hydrolysis of tetraethylorthosilicate (TEOS) using the same microemulsion at ambient temperature for 72h using NH_4OH as a catalyst. TEM micrographs showed that the silicon oxide particles were in the size around 20 nm with a very uniform particle size distribution (Fig. 6a) and they are amorphous. Silica coated iron oxide particles prepared via microemulsion (Fig.6b) are 20-25nm with the thickness of the coating ~2-5nm and show much better refined core-shell structure compared with the micrographs reported [6]. X-ray diffraction pattern of the silica coated iron oxide nanoparticles displayed poor crystallinity as compared to the uncoated iron oxide nanoparticles even after hydrothermal treatment at 200°C for 5h. This may be attributed to the layer of silica coating restricting growth of iron oxide nanoparticles in hydrothermal.

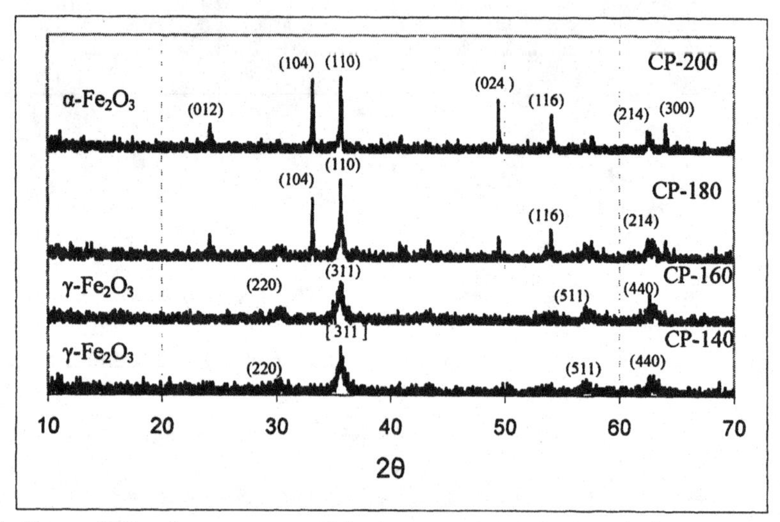

Fig. 5. X-ray diffraction patterns of the nanoparticles prepared via coprecipitation for the hydrothermal as a function of synthesis temperature for 5h.

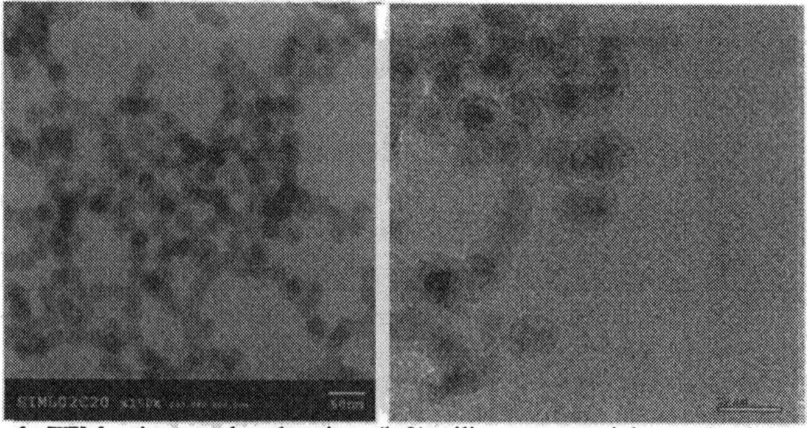

Fig. 6. TEM micrographs showing (left) silica nanoparticles and (right) iron oxide/silica nanopartiles prepared via microemulsions.

In order to further investigate the core-shell structure of the coated nanoparticles, the sample was examined using a High Resolution Transmission Electron Microscope equipped with a thin window Energy Dispersive X-ray Analysis. A nanoprobe was applied to the outer layer (shell) and the central part (core) of the nanoparticle. There was no iron element detected in the shell, indicating the iron oxide core being wrapped by a silicon oxide shell (Fig. 7).

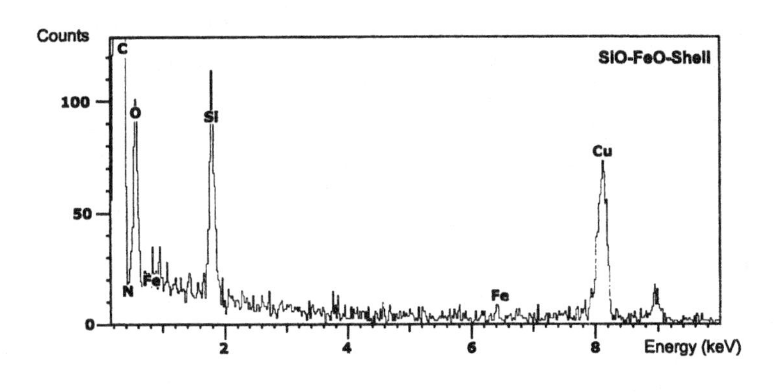

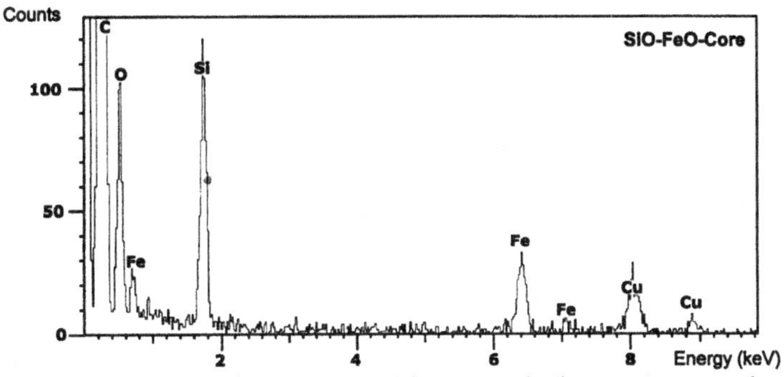

Fig. 7. EDX analysis of (top) outer layer and (bottom) central part of the nanoparticle.

CONCLUSIONS

Both microemulsion and coprecipitation methods were used to synthesize iron oxide nanoparticles. Hydrothermal post-treatment was also employed to enhance the crystallinity of the iron oxide nanoparticles. The iron oxide nanoparticles produced via both microemulsion and coprecipitation revealed increasing crystallinity after hydrothermal treatment. A more stable phase (α-Fe_2O_3) was formed for the nanoparticles prepared via microemulsions, and a metastable phase (γ-Fe_2O_3) was observed for the nanoparticles prepared via coprecipitation after hydrothermal treatment at 140°C for 5h. Spherical silica nanoparticles were also synthesized by hydrolysis of TEOS in the microemulsion. TEM micrographs showed that iron oxide nanoparticles were successfully coated by silica via microemulsions: dark spots (iron oxide) being coated by a thin layer of light area (silica), which was further supported with EDX analysis.

REFERENCES

[1]Y. Haik, V. Pai, and C. Chen, "Development of magnetic device for cell separation", *Journal of Magnetism and Magnetic Materials*, **194** 254-261 (1999).

[2]P. Tartaj and C.J. Serna, "Microemulsion-assisted synthesis of tunable superparamagnetic composites", *Chemistry of Materials*, **14** 4396-4402 (2002).

[3]I. Safarik and M. Safarikova, "Use of magnetic technique for the isolation of cells", *Journal of Chromatography B*, **722** 33-53 (1999).

[4]J.I. Martin, J. Nogues, K. Liu, J.L. Vicent, I.K. Schuller, "Ordered magnetic nanostructures: fabrication and properties", *Journal of Magnetism and Magnetic Materials*, **256** 449-501 (2003).

[5]T. Hyeon, S.S. Lee, J. Park, Y. Chung and H.B. Na, "Synthesis of high crystalline and monodispese maghemite nanocrystallines without a size-selection process", *J. Amer. Chem. Soc.,* 123 12798-12801 (2001).

[6]S. Swadeshmukul, T. Rovelyn., T. Nikoleta, D. Jon, H. Arthur, and T. Weihong, "Synthesis and Characterization of Silica-Coated Iron Oxide Nanoparticles in Microemulsion", *Langmuir*, **17** 2900-2906 (2001).

[7]F. Caruso, "Nanoengineering of particle surfaces", *Advanced Materials*, **13** 11-22 (2001).

[8]C.J. Zhong and M.M. Maye, "Core-shell assembled nanoparticles as catalysts", *Advanced Materials*, **13** 1507-11 (2001).

[9]F. Caruso, M. Spasova, A. Susha, M. Giersig and R.A. Caruso, "Magnetic nanocomposite particles and hollow spheres constructed by a sequential layering approach", *Chemistry of Materials*, **13** 109-116 (2001).

[10]L. Levy, Y. Sahoo, K.S. Kim, E.J. Bergey and P.N. Prasad, "Nanochemistry: synthesis and characterisation of multifunctional nanoclinics for biological applications", *Chemistry of Materials*, **14** 3715-21 (2002).

[11]J. Shi, S. Gider, K. Babcock and D.D. Awschalom, "Magnetic Clusters in Molecular Beams", *Metals and Semiconductors Science*, **271** 937-41 (1996).

[12]M. Boutonnet, J. Kizling, P. Stenius, and G. Marie, "The Preparation of Monodisperse Colloidal Metal Particles from Microemulsions", *Colloid Surface* **5** 209-225 (1982).

ACKKNOWLEDGEMENT

The authors would like to thank British Royal Society and Loughborough University for supporting this project.

Characterization and Properties of Nanomaterials

SYNTHESIS OF Si-BASED NANOWIRES

K. Saulig-Wenger, D. Cornu, F. Chassagneux, S. Parola and P. Miele
Laboratoire Multimatériaux et Interfaces, UMR CNRS 5615
Université Claude Bernard – Lyon 1
43 Bd du 11 Novembre 1918, F-69622 Villeurbanne Cedex, France

T. Epicier
GEMPPM UMR CNRS 5510
INSA Lyon
20 Avenue Albert Einstein, F-69621 Villeurbanne Cedex, France

ABSTRACT

Direct thermal treatment of a commercial silicon powder and graphite placed in an alumina crucible lead to cubic silicon carbide (β-SiC) and amorphous silicon dioxide (SiO_2) nanowires, depending on the atmosphere used during the pyrolysis process. The structure and composition of these nanowires have been investigated by the means of HRTEM, EDX and EELS.

INTRODUCTION

Since the discovery of carbon nanotubes (CNTs) in 1991 [1], numerous studies were devoted to nanoscale materials. Various nanoobjects such as nanotubes (WS_2 [2], BN [3-5]...) and nanorods (GaN [6], Si_3N_4 [7]...) have been prepared because they could be more efficient than CNTs for diverse applications (nanoelectronic, nanocomposites, etc.). Among these materials, silicon-based nanowires (NWs) such as SiC, SiO_2 or Si_3N_4 NWs could offer good properties for mechanical [8,9], electrical [10,11] or optical [12] applications. Several routes have been reported for the synthesis of those nanowires. The carbothermal reduction of silica with carbon nanocapsules [13] or carbon nanotubes (CNTs) [14] yields SiC NWs or SiO_2 sheathed SiC nanocables (NCs). Other routes using CNTs as templates yielded to Si_3N_4 NWs [7] or SiO_2 NWs [15]. Chemical vapor deposition (CVD) techniques [10] and reaction between CCl_4(g) and $SiCl_4$(g) [16] or Si powder in presence of sodium [17] have also been used to prepare SiC NWs.

However all those synthetic methods require either expensive starting materials such as CNTs or expensive equipment.

As an extend of the results previously described concerning the formation of β-SiC and h-BN coated β-SiC nanowires [18], we report in this paper a simple and inexpensive route to prepare silicon-based nanomaterials. Silicon carbide and silicon oxide NWs have been prepared and fully characterized by the means of HRTEM, EELS and EDX analyses.

EXPERIMENTAL

In a typical experiment, silicon powder (Aldrich 99.999%, 60 mesh) was placed in an alumina boat which contained piece of graphite. This boat was then placed in the alumina tube of an horizontal tubular furnace, the tube being previously degassed *in vacuo* before filling with the selected gas. Under the gas flow (5 $ml.min^{-1}$), the sample was heated up to the selected temperature, held for 1 hour then allowed to cool down to room temperature. Finally, a powder was scraped from the alumina boat and analyzed by the means of XRD, Scanning Electron Microscopy (SEM, Model N°S800, Hitachi), High-Resolution Transmission Electron Microscopy (HRTEM, Field Emission Gun microscope JEOL 2010F), Energy Dispersive X-ray spectroscopy (EDX, Link-Isis OXFORD analyzer) and Electron Energy-Loss Spectroscopy (EELS, Digi-PEELS GATAN).

RESULTS AND DISCUSSION

Direct pyrolysis up to 1200°C of a silicon powder and graphite placed in an alumina crucible yields various nanowires depending on the experimental conditions. The results obtained are summarized in Table I.

Table I. Nanowires obtained *vs.* experimental conditions

Gas	Temperature (°C)	Nanowires
N_2	1200	SiC
$Ar + O_2$	1200	SiO_2

1) β-SiC Nanowires

After thermal treatment under nitrogen up to 1200°C, β-SiC nanowires are formed in accordance with our previous results [18].

In fact, low resolution TEM observation revealed the presence of highly curved NWs (Fig. 1A). Their diameters fall in the range 5 - 60 nm. Due to their curvature and also to the presence of residual Si particles, the exact length of the NWs is difficult to evaluate but it can be estimate to be in the range 50 - 100 μm.

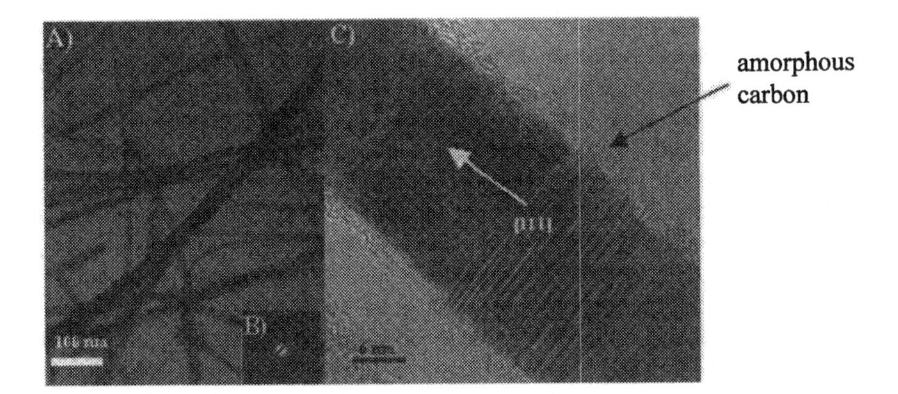

Fig. 1 : TEM images of β-SiC nanowires at low magnification (A) with a Selected Area Electron Diffraction (SAED) diagram (B) and HRTEM image of a β-SiC NW coated by amorphous carbon (C).

The chemical composition of the NWs has been investigated by EELS analysis (Fig. 2). On the spectrum, two main features at ~100 eV and ~284 eV are characteristic of Si-L and C-K edges, respectively. At ~532 eV, no feature is observed, pointing out the absence of oxygen in the NWs.

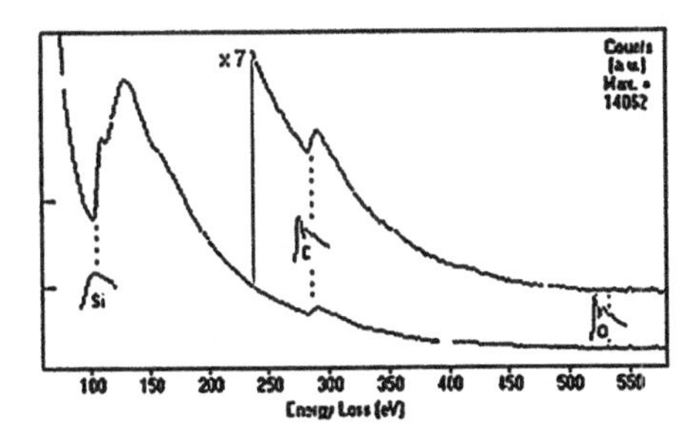

Fig. 2 : Typical EELS spectrum of a β-SiC nanowire

The Selected Area Electron Diffraction (SAED) (Fig 1B) shows that the nanowires are composed of the cubic polymorph of silicon carbide (β-SiC). The preferential growth direction is [111] and a high density of stacking faults normal to this direction has been observed.

According to the literature, three mechanisms can be envisaged for the growth of the nanowires : a screw dislocation, a VLS (Vapor-Liquid-Solid) or a VS (Vapor-Solid) growth process. No spiral NWs have been detected in the crude product so the screw dislocation process (which main feature is the formation of spiral NWs) is not involved. Moreover, the tips of the NWs have been studied and no globules have been observed but only few silicon nanoparticles were found. Since the silicon melting point is not reached during our experiment, these results suggest that the NWs grew *via* a VS mechanism. Therefore, carbon is dissolved into silicon particles at high temperature yielding super-saturated silicon and β-SiC nanowires grew up under cooling. This mechanism implies that the carbon is transported from graphite to the silicon powder. Since no traces of oxygen have been detected in the NWs, we assume that carbon is transported by nitrogen, *via* the formation of volatile carbon-nitrogen species as described earlier for carbonitride films formation [19]. In order to have experimental evidence of this assumption, a similar experiment was conducted under inert atmosphere. The thermal treatment up to 1200°C under argon of a silicon powder and graphite placed in an alumina boat did not yield nanowires.

It is interesting to notice that a coating has been observed on few NWs, as exemplified on fig. 1C. This coating is composed of amorphous carbon and is 2-4 nm thick. We can suggest that the formation of this amorphous coating occurred during the synthesis of the NW. We have no experimental evidence at the moment but further investigations have to be done in order to determine the mechanism of formation of this coating.

2) SiO_2 nanowires

After thermal treatment of the silicon powder under a mixture of argon and oxygen, a white product is collected on the top of the alumina crucible. SEM observations show that the sample is composed of numerous nanowires, without residual particles (Fig. 3A). The NWs are highly aligned along the gas flow direction and their length is estimate roughly around 100 μm. Their diameters are in the range 5 - 300 nm with an average diameter of 60 nm.

Electron diffraction and HRTEM image of the NW indicate that they are amorphous (Fig. 3B). As exemplified in fig. 3C, EELS analysis performed with a 2 nm electron probe reveals two main features at ~100 eV and ~532 eV corresponding to the Si-L edge and the O-K edge, respectively. A quantitative analysis by EELS indicate the following ratio : [Si]/[O] = 0.5 +/- 0.02. According to these results, amorphous SiO_2 nanowires are formed. It is also interesting to notice that no globules or particles have been detected at the tips of the nanowires.

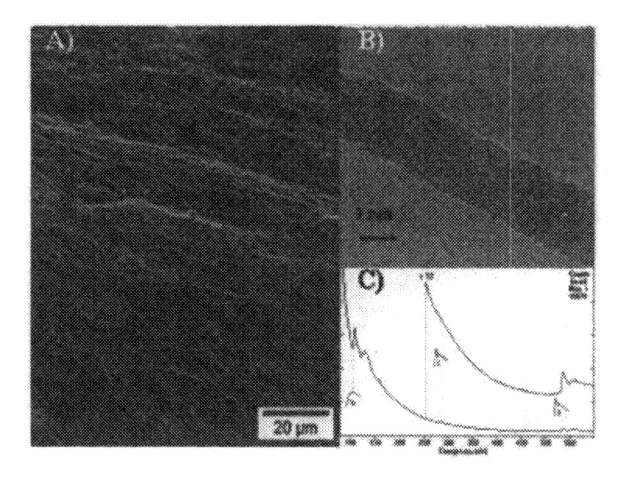

Fig. 3 : (A) SEM image of the highly aligned NWs. HRTEM image of an amorphous SiO_2 nanowire (B) and the corresponding EELS spectrum (C)

According to the literature, Zhang *et al.* have obtained a wide range of nanowires and particularly similar amorphous SiO_2 nanowires by heating silicon (or a mixture of silicon and silica) under argon containing residual oxygen [20]. The authors proposed that the growth process of the nanowires is mainly a VS mechanism, involving the formation of SiO(g) [20]. On the basis of their results, some experiments were conducted without a piece of graphite in the alumina boat. We observed that a direct thermal treatment of a silicon powder placed in an alumina boat did not allow the formation of SiO_2 nanowires with our experimental conditions. Moreover, under the same experimental conditions, when a piece of graphite was added in the alumina boat, numerous nanowires were formed. This result evidenced the role of graphite in the growth of these nanowires. We assume that the reaction of oxygen with carbon produced CO_2(g) (Eq. 1) and that it is the reaction of the latter with silicon which yield SiO(g) (Eq. 2).

$$C\,(s) + O_2\,(g) \rightarrow CO_2\,(g) \quad (1)$$

$$CO_2\,(g) + Si\,(s) \rightarrow CO\,(g) + SiO(g) \quad (2)$$

Further investigations are in progress in order to confirm these assumptions and the role of graphite in the growth of these nanowires.

CONCLUSION

Si-based NWs have been obtained by a simple method which consists in the pyrolysis at 1200°C of a silicon powder under different atmosphere. No catalysts were used and a VS process has been proposed as the growth mechanism for both β-SiC and amorphous SiO_2 nanowires. Some works are still in progress in order to have experimental evidence for these proposed growth mechanisms and the next step of this study will also involve the measurement of the physical (mechanical, electrical, etc.) properties of these nanowires.

ACKNOWLEDGEMENT

We gratefully thank the CLYME (Consortium Lyonnais de Microscopie Electronique) for the access to the TEM-FEG microscope.

REFERENCES

[1] S. Iijima, «Helical microtubules of graphitic carbon», *Nature*, **354** 56-58 (1991).

[2] R. Tenne, L. Margulis, M. Genut, G. Hodes, «Polyhedral and Cylindral structures of tungsten disulphide», *Nature*, **360** 444-446 (1992).

[3] N. G. Chopra, R. J. Luyken, K. Cherrey, V. H. Crespi, M. L. Cohen, S. G. Louie, A. Zettl, «Boron nitride nanotubes», *Science*, **269** 966-967 (1995).

[4] W. Han, Y. Bando, K. Kurashima, T. Sato, «Synthesis of boron nitride nanotubes from carbon nanotubes by a substitution reaction», *Applied Physics Letters*, **73** [21] 3085-3087 (1998).

[5] D. Golberg, Y. Bando, K. Kurashima, T. Sato, «Synthesis, HRTEM and electron diffraction studies of B/N-doped C and BN nanotubes», *Diamond and Related Materials*, **10** 63-67 (2001).

[6] W. Han, S. Fan, Q. Li, Y. Hu, «Synthesis of Gallium Nitride Nanorods Through a Carbon Nanotube-Confined Reaction», *Science*, **277** 1287-1289 (1997).

[7] W. Han, S. Fan, Q. Li, B. Gu, X. Zhang, D. Yu, «Synthesis of silicon nitride nanorods using carbon nanotubes as a templates», *Applied Physics Letters* **71** [16] 2271-2273 (1997).

[8] E.W. Wong, P.E. Sheehan, C.M. Lieber, «Nanobeam mechanics : elasticity, strenght and toughness of nanorods and nanotubes», *Science*, **277** 1971-1975 (1997).

[9] Y. Zhang, N.Wang, R. He, Q. Zhang, J. Zhu, Y. Yan, «Reversible bending of Si_3N_4 nanowire», *Journal of Materials Research*, **15** [5] 1048-1051 (2000).

[10] K. W. Wong, X. T. Zhou, F. C. K. Au, H. L. Lai, C. S. Lee, S. T. Lee, «Field-emission characteristics of SiC nanowires prepared by chemical-vapor deposition», *Applied Physics Letters*, **75** [19] 2918-2920 (1999).

[11] Z. Pan, H.-L. Lai, F. C. K. Au, X. Duan, W. Zhou, W. Shi, N. Wang, C.-S. Lee, N.-B. Wong, S.-. Lee, S. Xie, «Oriented silicon carbide nanowires : synthesis and field emission properties», *Advanced Materials*, **12** [16] 1186-1190 (2000).

[12] D.P. Yu, L. Hang, Y. Ding, H.Z. Zhang, Z.G. Bai, J.J. Wang, Y.H. Zou, W. Qian, G.C. Xiong, S.Q. Feng, «Amorphous silica nanowires : Intensive blue light emitters», *Applied Physics Letters*, **73** [21] 3076-3078 (1998).

[13] Y. H. Gao, Y. Bando, K. Kurashima, T. Sato, «The microstructural analysis of SiC nanorods synthesized through carbothermal reduction», *Scripta Materialia*, **44** 1941-1944 (2001).
[14] C. C. Tang, S. S. Fan, H. Y. Dang, J. H. Zhao, C. Zhang, P. Li, Q. Gu, «Growth of SiC nanorods prepared by carbon nanotubes-confined reaction», *Journal of Crystal Growth*, **210** 595-599 (2000).
[15] B.C. Satishkumar, A. Govindaraj, E.M. Vogl, L. Basumallick, C.N.R. Rao, «Oxide nanotubes prepared using carbon nanotubes as templates», *Journal of Material Research*, **12** [3] 604-606 (1997).
[16] Y. Zhang, N. Wang, R. He, X. Chen, J. Zhu, «Synthesis of SiC nanorods using floating catalyst», *Solid State Communications*, **118** 595-598 (2001).
[17] Q. Lu, J. Hu, K. Tang, Y. Qian, G. Zhou, X. Liu, J. Zhu, «Growth of SiC nanorods at low temperature», *Applied Physics Letters*, **75** [4] 507-509 (1999).
[18] K. Saulig-Wenger, D. Cornu, F. Chassagneux, G. Ferro, T. Epicier, P. Miele, «Direct Synthesis of β-Sic and h-BN coated β -SiC nanowires», *Solid State Communications*, **124** 157-161 (2002).
[19] C. Popov, M.F. Plass, R. Kassing, W. Kulisch, «Plasma chemical vapor deposition of thin carbon nitride films utilizing transport reactions», *Thin Solid Films*, **355-356** 406-411 (1999).
[20] Y. Zhang, N. Wang, S. Gao, R. He, S. Miao, J. Liu, J. Zhu, X. Zhang, «A simple method to synthesize nanowires», *Chemistry of Materials*, **14** [8] 3564-3568 (2002).

CHARACTERIZATION OF NANOMETER-SCALE COLUMNAR AND LOW-DENSITY BOUNDARY NETWORK STRUCTURES IN HYDROGENATED AMORPHOUS CARBON FILMS

Eiji Iwamura
PRESTO, Japan Science and Technology Corporation
RCAST, The University of Tokyo
4-6-1 Komaba, Meguro-ku
Tokyo 153-8904, Japan
e-mail: iwamura@odin.hpm.rcast.u-tokyo.ac.jp

ABSTRACT

The nano-scale network structures consisting of columns and inter-column regions were examined in hydrogenated amorphous carbon films using energy-filtered transmission electron microscopy and scanning probe microscopy. Amorphous carbon films with 0.2-1.2 μm in thickness were sputter-deposited on Si wafers at various $Ar+CH_4$ gas pressures. The width of the inter-column region increased from less than 1 nm to 20 nm, as the sputter gas pressure increased from 0.27 Pa to 4 Pa. The inter-column region appeared not to be porous but densely filled with carbon atoms. The valence electron density of the inter-column region was found to be at least 30% lower than that of the column region which contained graphite-like clusters. Electrical conductivity was detected in the inter-column region, while the column region was insulating.

INTRODUCTION

The columnar structures commonly found in vapor-deposited thin films have been classified by what have been termed structure zone models.[1] It is reported that amorphous films as well as polycrystalline films form zone 1 and zone T structures under deposition conditions of low adatom mobility.[2-4] The microstructure in amorphous thin films can be characterized by columns with relatively high density surrounded by lower density regions. The lower density regions usually exhibit network structures, which have been called void networks, crack networks or honeycomb-like networks.[5] It is likely that such an inhomogeneous but a quasi-periodic and an anisotropic structure plays an important role in film properties. The column/void network structure has recently attracted much interest as it can be used in nano- and micro-scale channels for fluidic transportation.[6] Furthermore, it is suggested that conducting channels extending from the film surface affect field emission properties in amorphous carbon films.[7] The evolution of the columnar structures and geometry of surface morphology in amorphous films have been investigated since late 70's, however, the characteristics of the lower density region are not fully understood.

The purpose of this study is to provide a detailed structural analysis of the nano-scale networks formed in sputter-deposited hydrogenated amorphous carbon films. The film structures depending on sputtering gas pressure were examined in terms of the dominant structural unit size, the difference of density between columns and the surrounding regions, and their electric properties.

EXPERIMENTAL

Hydrogenated amorphous carbon (a-C:H) thin films were prepared by dc magnetron sputtering. Using an anisotropic graphite target, a-C:H films with 0.2-1.2 μm in thickness were deposited onto <100>Si substrates in $Ar+CH_4$ gas mixtures at a pressure range from 0.27 to 4.0 Pa. Hydrogen content in the films was in the range of 40±5at% as measured by Elastic Recoil Detection Analysis (ERDA). Neither intentional heating nor bias voltage were applied to the substrates. The substrate temperature during deposition was estimated to be up to 473 K. The base chamber pressure was less than 2.6×10^{-3} Pa. The sputtering dc power density was 0.085 $W\cdot mm^{-2}$. The resultant sputtering rate changed from 2.4 to 17 $nm\cdot min^{-1}$ as the gas pressure increased.

TEM experiments were performed using a HITACHI HF-2000 operating at 200kV equipped with a Gatan Model 678 Imaging Filter. Electron energy loss spectra were acquired in the energy range up to 38 eV. The energy resolution estimated from a full width at half maximum of a zero-loss peak was 0.6 eV. Zero-loss and plasmon-loss images were obtained using an energy slit which was set to be 3.0 eV and centered at 0, 16, 19, 22, 25, and 28, respectively.

Topographic and electric current images on the top of film surface were obtained using a scanning probe microscopy (SPM-9500J3, Shimadzu Corporation). The measurements were performed in contact mode using an n^+-Si/$PtIr_5$ tip (POINTPROBE). Negative bias voltage up to 1 V was applied to the samples for electron current imaging.

RESULTS AND DISCUSSION

Figures 1(a)-1(d) show plan-view TEM microstructures of a film deposited at a gas pressure of 0.27, 2.0 and 4.0 Pa, respectively. The contrast seen in the figures is resulted from a local change in film density. The light region which formed a network structure is described as a density deficient region. As the gas pressure increased, the width of the light region increased from less than 1nm to about 20 nm. Discontinuity of film structures was not recognized, while only a small number of voids or microporosities appeared to be in the light region.

Figures 2(a)-2(c) show cross-sectional TEM micrographs showing the evolution of film structures. Columnar structures seen as dark region developed from the bottom of the film to the top surface increasing column size. Defining the size of dominant surface morphology and column size in the vicinity of the top surface as $d_{surface}$ and d_{column}, respectively, the relationship between the structural unit sizes and film thickness is shown in Fig. 3. The $d_{surface}$ was measured by SEM and SPM, and the d_{column} by TEM micrographs. Although those plots included the data from films deposited at various gas pressures, the dominant structural unit sizes showed linear relationship to film thickness. Each unit size increased linearly with the 0.8 power of the film thickness, independent of gas pressures. The column sizes were 1/3 of the sizes of the dominant surface morphology. It is reported that the dominant film structures evolve with film thickness according to a power low relation.[8,9] As those relationship between the dominant structural units size and film thickness are the same as previously noted by Messier et al,[8]

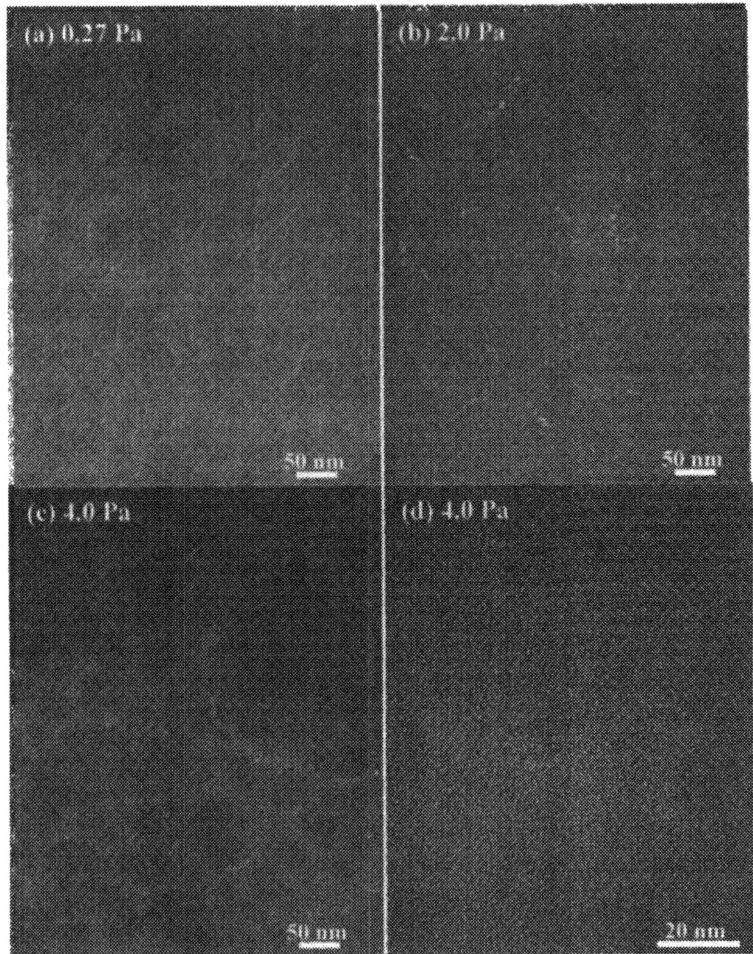

Fig. 1. Plan-view TEM micrographs of a-C:H films. Film thicknesses are (a) 430 nm, (b) 1200 nm, and (c),(d) 680 nm, respectively.

the evolution of main frame of film structures is presumably dominated by the low thermal mobility of adatoms and shadowing effects during sputtering as well as a-Ge and a-Si:H.[3,8,9] On the other hand, higher sputtering gas pressures lead to reduce adatom mobility and randomize the direction of sputtering atoms. As the results, the inter-column region tends to contain voids and microporosities since shadowing effects and/or surface diffusion dominate the evolution of film structures. However, most of the inter-column region appeared to be densely filled even in the film with the relatively wide inter-column spacing as shown in Fig.1 (d). It is likely that bombardment or chemically induced mobility caused to fill the space surrounding the columns.

Figures 4(a)-4(c) show electron diffraction patterns taken from the column region deposited at various gas pressures. Broad diffusion rings corresponding to d=0.12 and 0.21 nm which were generally observed in a-C films, were identified

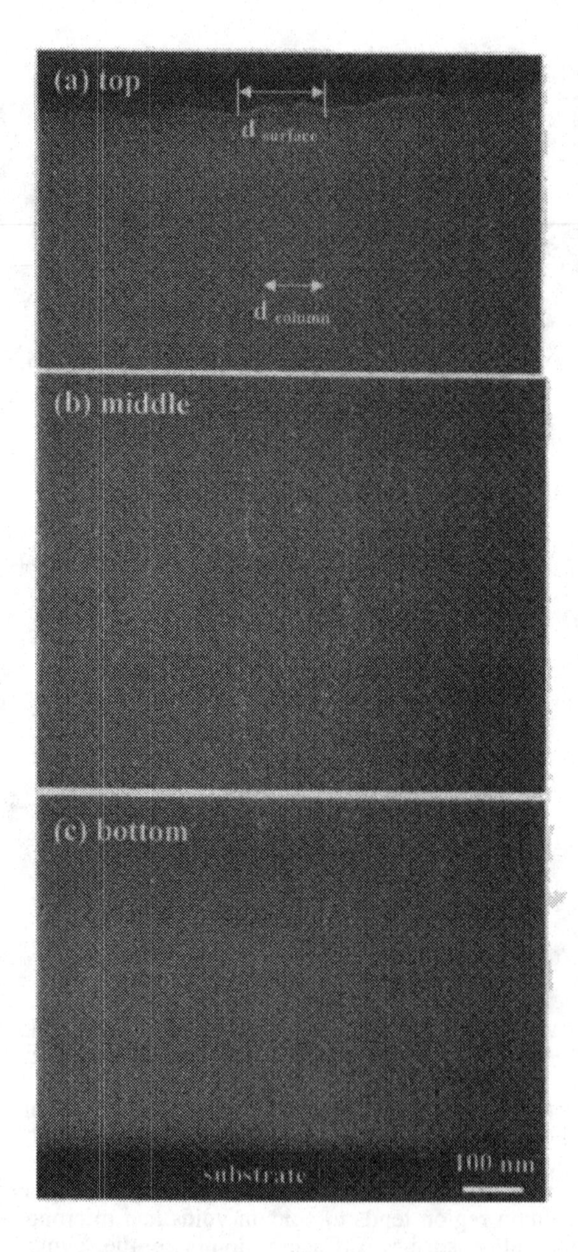

Fig.2. Cross-sectional TEM micrographs of an a-C:H film deposited at 2 Pa.

in all of the films. Diffraction from c-plane of graphite structures was indistinct and no texture was recognized. The fact that an additional ring corresponding to d=0.16 nm was observed in the films deposited at 4 Pa indicated that a certain structural ordering developed by decreasing adatom mobility.

Figure 5 shows EELS spectra in plasmon-loss energy region obtained from a

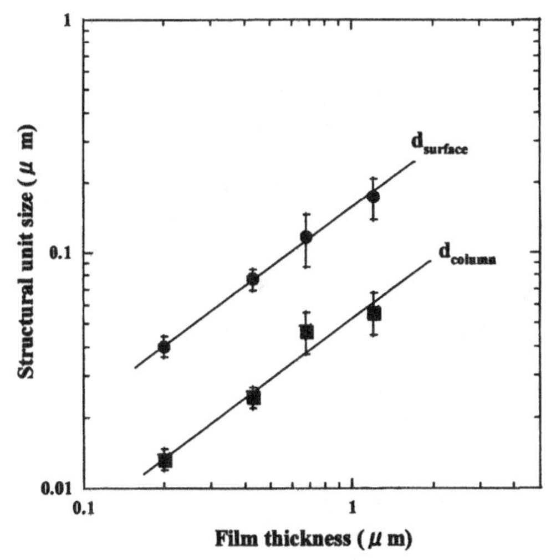

Fig. 3. The dependence of the dominant surface morphology size ($d_{surface}$) and column size (d_{column}) on film thickness.

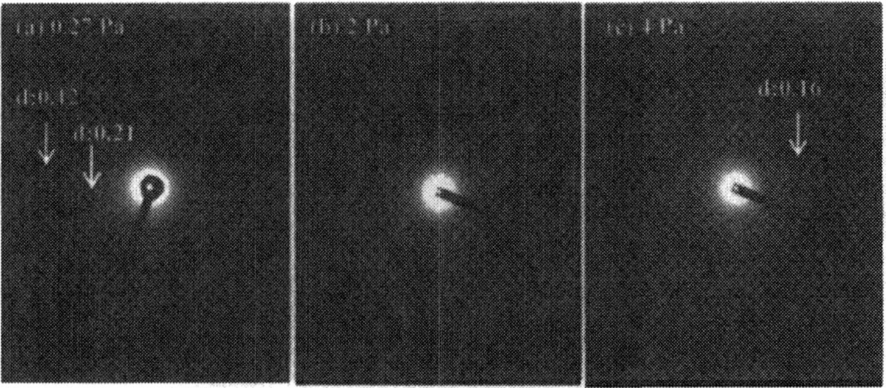

Fig. 4. Electron diffraction patterns of the column in the a-C:H films deposited at (a) 0.27 Pa, (b) 2 Pa, and (c) 4 Pa, respectively.

column and an inter-column region. The π and $\pi+\sigma$ plasmon-loss peaks from the column region were measured as about 5 eV and 23 eV, respectively. The apparent peak of the π resonance indicated that the column region contained graphite-like cluster structure, although the cluster structure was not identified in the TEM images and the electron diffraction patterns. On the other hand, the π resonance peak became indistinct in the spectrum from the inter-column region.

Figures 6(a)-6(f) show energy-filtered images showing the difference of plasmon energy between the column and the inter-column region. The zero-loss image shows the contrast which can be attributed to a local fluctuation of atomic density. The inter-column region was distinguishable as seen in the light network structures with 1-2 nm in width (indicated with arrows in Fig.6a). The density of the inter-column region deducing from the difference of spectrum intensities was

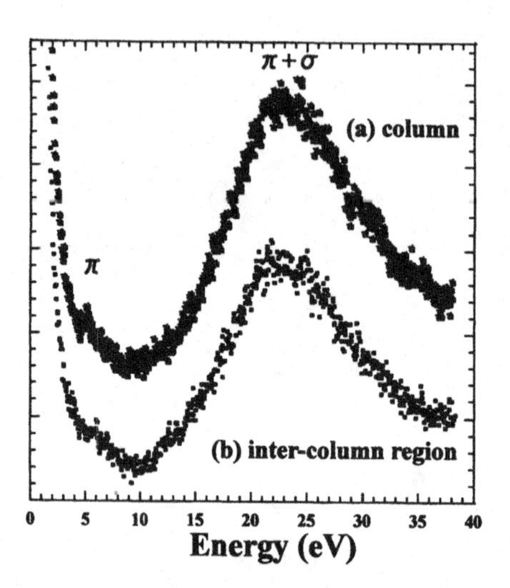

Fig.5. Electron energy loss spectra from (a) the column region and (b) the inter-column region in an a-C:H film deposited at 2 Pa.

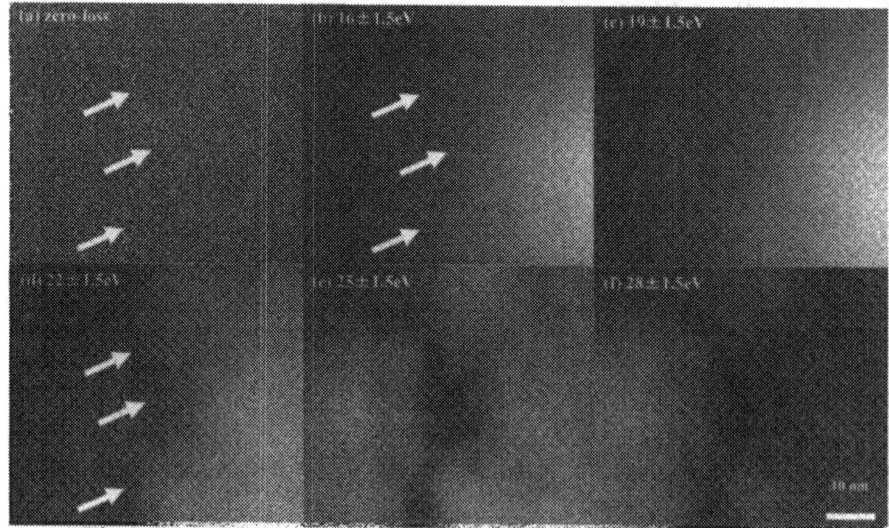

Fig.6. Energy-filtered plan-view TEM micrographs of an a-C:H film deposited at 2 Pa.

about 12 % less than that of the column region. The inter-column region showed brighter contrast in the energy range of 16±1.5 eV (Fig.6b) and became indistinct relative to that of the columns in the range of 19±1.5 eV (Fig.6c). The inter-column region exhibited dark contrast in the images of the higher energy range (Fig.6d-f). Therefore, the inter-column region had plasmon-loss energy (Ep) in the range less than about 19 eV. On the other hand, the column region showed the brightest contrast in the range of 25±1.5 eV, as expected from the EELS spectrum shown in Fig.5. The energy of the π+σ plasmon-loss peak is related to the valence electron density using the Drude equation:

$$n_e=mEp^2/4\pi e^2,$$

where n_e is the valence electron density, m is the electron mass, and e is the electronic charge. Assuming the Ep of the inter-column region is 19 eV and that of the column region is 23 eV, the valence electron density in the inter-column region is about 32 % at least less than that in the column region. The difference was much larger than that in atomic density estimated from the intensity of zero-loss peak.

It is expected that the difference of the density influences mechanical, optical or electrical properties of the films. Figure 7 shows a topographic image superimposed with an electric current image (bright dots in the figure). The region seen as light corresponds to the column region. The difference of height between column-top and bottom of the surrounding region was about 1 nm. Conductivity with about 30-40 nA was detected in the inter-column region, while the column region was insulated. The fact indicates that the lower density leads to increase of anti-bonding electrons in amorphous carbon network in the inter-column region.

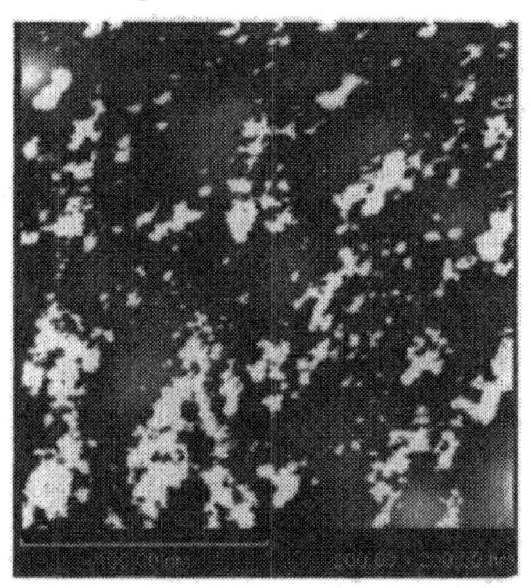

Fig. 7. Topographic SPM image superimposed with electric current image of the top surface of an a-C:H film deposited at 4 Pa.

SUMMARY

Nano-scale network structures consisting of columns and inter-column regions in sputter-deposited hydrogenated amorphous carbon thin films were examined using energy-filtered TEM and scanning probe microscopy. The following results were obtained.

(1) As pressure of process gas increased to 4 Pa, the width of the inter-column region increased up to 20 nm. Discontinuity of film structures was not observed and the inter-column region appeared to be densely filled rather than

porous.

(2) The size of the dominant structure in surface morphology and microstructure in the vicinity of the top surface was strongly depended on film thickness rather than sputtering gas pressure. The dominant surface morphology size was nearly linear with the 0.8 power of the film thickness, and the dominant column size was 1/3 of the size of the surface morphology.

(3) Energy filtering imaging technique revealed that the valence electron density in the inter-column region had at least 30% lower than that in the column region.

(4) The inter-column region showed electrical conductivity while the column region was insulated. The difference of conductivity in the film structures is presumably attributed to increase of anti-bonding electrons in the inter-column region.

ACKNOWREDGEMENTS

The author would like to thank Mr. H. Nomura of Matsuyama Giken Co, Ltd., for providing the sputter-deposited a-C:H films. Thanks are also due to Mr. M. Matsuda of Shimadzu Corporation for the SPM measurement.

REFERENCES

[1] J.A. Thornton, "High Rate Thick Film Growth," *Ann. Rev. Mater. Sci.*, **7** 239-60 (1977).

[2] R. Messier and R.C. Ross, "Evolution of Microstructure in Amorphous Hydrogenated Silicon," *J. Appl. Phys.*, **53** [9] 6220-25 (1982).

[3] R. Messier, A.P. Giri and R.A. Roy, "Revised Structure Zone Model for Thin Film Physical Structure," *J. Vac. Sci. Technol.*, **A2** [2] 500-03 (1984).

[4] G.S. Bales and A.Zangwill, "Macroscopic Model for Columnar Growth of Amorphous Films by Sputter Deposition," *J. Vac. Sci. Technol.*, **A9** [1] 145-49 (1991).

[5] A. Staudinger and S. Nakahara, "The Structure of the Crack Network in Amorphous Films," *Thin Solid Films*, **45** 125-33 (1977)

[6] W.J. Nam, S. Bae, A.K. Kalkan and S.J. Fonash, "Nano- and Microchannel Fabrication Using Column/void Network Deposited Silicon," *J. Vac. Sci. Technol.*, **A19** [4] 1229-33 (2001).

[7] J. Robertson, "Diamond-like amorphous carbon," *Mat. Sci. Eng.*, **R37** [4-6] 129-281 (2002).

[8] R. Messier and J.E. Yehoda, "Geometry of thin-film morphology," *J. Appl. Phys.*, **58** [10] 3739-46 (1985).

[9] R. Messier, "Toward Quantification of Thin Film Morphology," *J. Vac. Sci. Technol.*, **A4** [3] 490-95 (1986).

PROPERTIES OF TRANSPARENT CONDUCTING COATINGS (TCO) MADE BY CHEMICAL NANOTECHNOLOGY PROCESS

Naji Al-Dahoudi, Ahmed Solieman and Michel A. Aegerter*
Institut fuer Neue Materialien gGmbH - INM
Im Stadtwald, Geb. 43
66123 Saarbruecken / Germany

ABSTRACT

Suspensions made of crystalline In_2O_3:Sn (ITO) nanoparticles with average size smaller than 25 nm redispersed in adequate organic compounds (solvent, binder, etc.) have been developed to obtain transparent thick conducting coatings using the dip, spin and spray processes. The coatings can be further processed and patterned by either a UV irradiation at room temperature or a thermal treatment up to 1000°C. Their optical (transmission, reflection, absorption), electrical (carrier density, mobility, resistivity) properties are reported. The low temperature UV treatment allows to coat glass and plastic substrates with a typical specific resistivity of 40×10^{-3} ohm cm, comparable to that of commercial organic conducting polymers, while that of the heat treated coatings can be as low as 1×10^{-3} ohm cm. Their mechanical properties (adhesion, scratch resistance and hardness) determined using various DIN tests are also reported.

INTRODUCTION

Transparent conducting oxide (TCO) coatings are today essential components in numerous applications for which a combination of high visible transmission and a high electrical conductivity is required (electrodes in opto-electronic devices, IR reflecting or heatable transparent substrates, electromagnetic shielding, static dissipators, etc.).

* corresponding author: phone: +49-681-9300-317, fax: +49-681-9300-249, email: aegerter@inm-gmbh.de

Wide band gap ($E_g > 3$ eV) n-type semiconductors such as Sn doped indium oxide (ITO), Sb or F doped tin oxide (ATO, FTO) are the most often used inorganic TCO materials. Such coatings are commercially available and are usually deposited by rather costly techniques such as Physical Vapor Deposition (PVD) or Chemical Vapor Deposition (CVD) with sheet resistance - the ratio of the resistivity ρ to the thickness t of the coating - $R_□ < 10\ \Omega_□$ and adequate optical properties.

Wet chemical processing - sol-gel and metal-organic deposition (MOD) - are low cost industrial alternatives if a low sheet resistance is not of prime importance. They allow to coat small to large flat substrates but also complex shaped substrates and cavities with excellent homogeneity and better smoothness, a difficult task with PVD and CVD processes [1,2]. However, both techniques require heat treatment steps at high temperature and only allow to coat rather thin coatings.

The paper describes the preparation of ITO coatings using a nanotechnology approach in which thick single layers ($t \leq 1$ μm) can be deposited on glass and especially plastic substrates using sols made by redispersing conducting crystalline nanoparticles in an adequate suspension. The process separates the crystallization step of the TCO material from the film formation.

EXPERIMENTAL

The preparation of ITO powders, paste and sols is reported in [3]. Typically, a highly stable (> 1 year) blue paste of ITO nanoparticles (≤ 25 nm) with monomodal size distribution and solid content up to 75 vol% has been obtained without evidence of agglomeration if a pH < 6 is maintained. Adequate sols to coat substrates to be sintered at T > 300°C are obtained by diluting the paste in ethanol (typically 25 wt% of solid content). However the major advantage of the concept is that coupling agents (binders) like 3-Methacryloxypropyltrimethoxysilane (MPTS) can be used to link the particles together through UV irradiation so that nanocomposite coatings with more than 95 wt% of ceramic materials can be obtained on heat sensitive substrates (fig. 1).

COATING SINTERED AT HIGH TEMPERATURE

Figure 2 shows the electrical parameters (resistivity ρ, carrier concentration N and mobility μ) of pure nanoparticle ITO coatings (Sn/In = 7/93 in mole%) about 500 nm thick deposited on a fused quartz plate sintered in air up to 1000°C, post annealed in forming gas (FG: N_2/H_2:92/8) at 350°C /30 min and stored in air during 7 days. The resistivity decreases gradually by increasing the firing temperature down to 1.2×10^{-2} Ωcm at T = 1000°C. The behavior reflects a better sintering and the growth of the particles that improve essentially the electron mobility (decrease of the grain boundary scattering). A further improvement is obtained by post annealing the coatings. This process has almost no influence on the mobility but drastically increases the electron concentration N. When the coatings are left in vacuum or a protective atmosphere the values of ρ and N remain stable but they unfortunately change with time when stored in air. The variation depends on the sintering temperature and is probably due to the back diffusion into the coatings of oxygen species acting as electron traps. The effect is more pronounced for coatings sintered at 550°C (porosity of 52%) and almost negligible for coatings sintered at 1000°C (porosity of 37%).

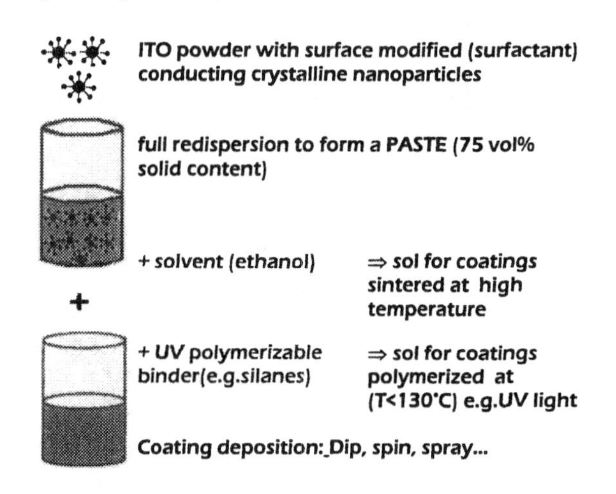

Figure 1. Procedure to get sols for wet-coating deposition.

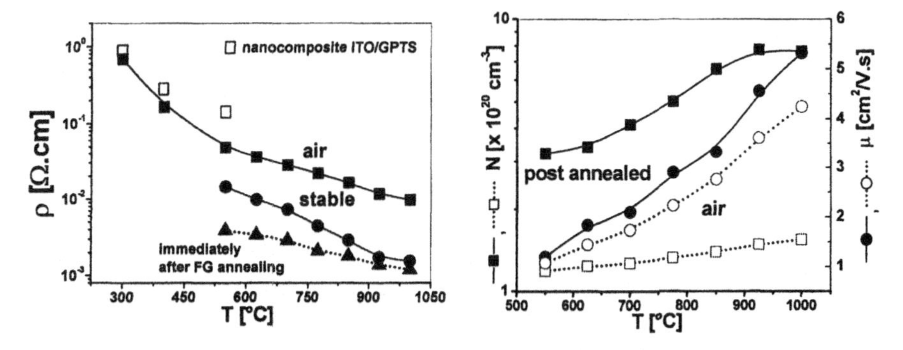

Figure 2. Left: Specific resistivity vs. sintering temperature of nanoparticles ITO coatings sintered in air (■), after post annealing in forming gas (FG) (▲) after 7 days storage in air (●). Right: electron concentration N (■,□) and mobility μ (●,○). t ≈ 500 nm.

Typical optical properties (up to $\lambda = 3$ μm) are shown in figure 3 for a coating sintered at 550°C ($\rho = 4.8 \times 10^{-2} \Omega cm$) and further reduced in FG ($\rho = 1.5 \times 10^{-2} \Omega cm$). Both films exhibit a high transmission of about 90% in the visible range. The coating sintered in air shows a larger transmission window extending from the UV up to λ= 2.3 μm (T = 50%) than the post annealed one (λ= 1.25 μm). This originates from the presence of a high broad absorption band related to the free carriers. This photon loss is known to be proportional to Nt/μ. The value of A is particularly high for the postannealed coatings, as t and N are high and μ is small.

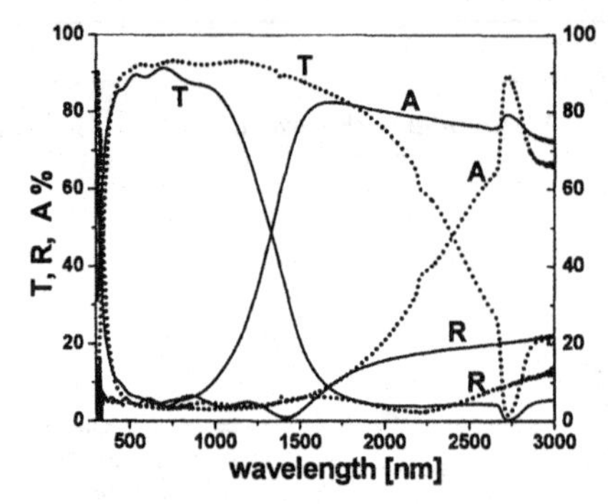

Figure 3. Optical transmission T, reflection R and absorption A=1-T-R of a pure nanoparticle ITO coating sintered at 550°C and FG annealed. t ≈ 500 nm . Dashed lines: air sintering; solid lines: post-annealed in forming gas.

The far infrared reflectance depends on the dc resistivity and increases when ρ decreases (Figure 4). The results are only in partial agreement with the Drude theory and only the low resistivity samples follow the Hagen-Ruben relation $1\text{-}R \propto \rho^{1/2}$. This indicates that the condition $\omega << e/m^{*}\mu$ (inverse of the frequency independent average electron relaxation time) is only satisfied for low resistivity coatings.

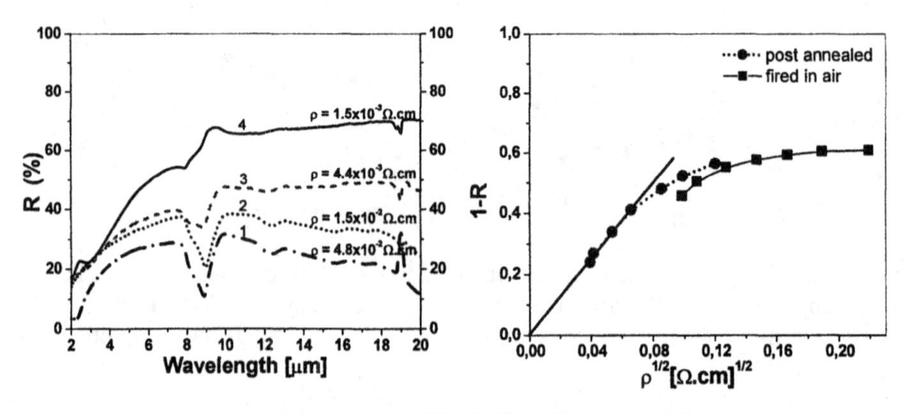

Figure 4. Left: IR optical reflection of ITO films fired at different sintering temperatures T_S and annealed in forming gas at 350 °C: 1. T_s = 550 °C fired in air, 2. T_s = 550 °C and annealed, 3. T_s = 775 °C and annealed, 4. T_s = 1000 °C and annealed. Right: 1-R vs $\rho^{1/2}$ for reflection data taken at λ = 22 μm.

The mechanical properties are shown in table 1. The properties become better as the firing temperature is increased but are not very good in view of some use. However, the addition of a small amount of a prehydrolyzed organosilane drastically improves the results (see below) without practically modifying the values of the resistivity (□, Fig. 2, left).

Table 1: Adhesion, abrasion resistance of sintered ITO films.

T (° C)	Tape test DIN 58196-K2	Cloth test DIN 5896-H25	Rubber test DIN 58196-G10
120	Totally removed	Class 5	Class 5
550	Partially removed	Class 3	Class 5
775	Ok	Class 2	Class 4
1000	Ok	Class 2	Class 3

COATING POLYMERIZED AT LOW TEMPERATURE

Coatings have been prepared with sols containing small amount of polymerizable prehydrolyzed organosilanes. Among them, the use of MPTS led to the best electrical, optical, structural and mechanical properties [4].

The lowest resistivity, $\rho = 4.5 \times 10^{-2}$ Ωcm, was obtained for coatings made with a sol containing 6 vol% MPTS, polymerized during 2 min under a 105 mW/cm² UV irradiation (Beltron) followed by a heat treatment at 130°C during 10 h and then annealed in N_2 atmosphere at 130°C during 2 h. This value is only stable if the coatings are kept in vacuum or protective atmosphere and it increases slightly to a stable value $\rho = 9.5 \times 10^{-2}$ Ωcm after 7 days in air. This variation is reversible if the samples are reexposed to UV light and is essentially due to the variation of the carrier density, N, and not the mobility μ (figure 5).

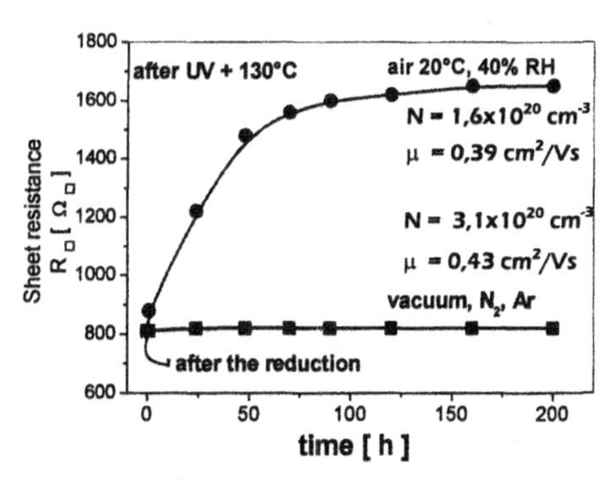

Figure 5. Time evolution of the resistivity of a 570 nm thick MPTS/ITO nanocomposite coating left in a protective atmosphere (vacuum, Ar, N_2) and in air. The coatings have been previously post annealed in a reducing atmosphere. N and μ were measured after a 200 h storage time.

The detail of the mechanism is not yet clear but it should involve oxygen species, adsorbed on the particle surface acting as free electron traps. The species are released by the UV irradiation, but when the coatings are kept in air, O_2 diffuses into the porous coating and are likely re-adsorbed on the surface of the ITO particles, decreasing N and increasing ρ. The ratio ρ(7 days)/ρ(t=0) depends on the thickness of the coating; it is 10 for t = 100 nm and 2 for t = 1 μm.

The surface morphology of the coatings observed by SEM (figure 6) consists of loosely packed globular grains (raspberry like) about 100 nm in size formed by the aggregation of the ITO nanoparticles linked together by a small strip (< 1 to 2 nm) of polymerized MPTS (darker regions). The coating roughness measured with a high resolution (AFM) on a 1 x 1 μm^2 area is Ra = 4 nm with a peak-to valley of 28 nm.

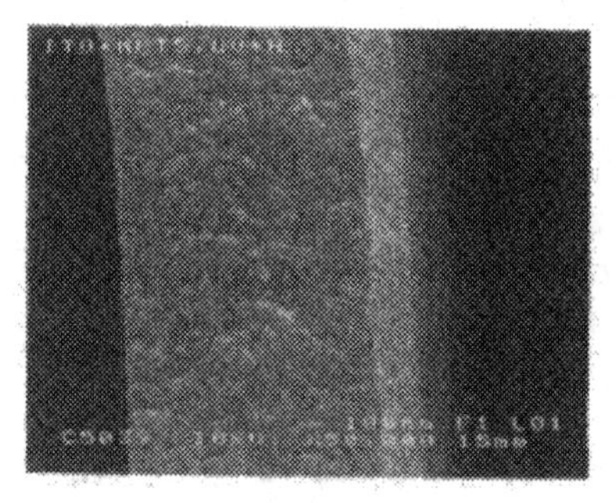

Figure 6. MPTS/ITO nanocomposite coating. Left: SEM picture of the surface. Right: HR-TEM cross-section.

The optical transmission and reflection of a coating deposited on a polycarbonate substrate is shown in figure 7. A high transmission of about 87% is observed in the visible range. The influence of the carriers is clearly seen by the strong absorption (A = 1 - T - R) occurring in the near IR range (900 < λ < 3 μm) and the increase of the reflection starting at $\lambda \approx$ 1.7 μm extending to R $\approx$ 45% at λ = 20 μm. Similar results have been obtained for coatings deposited on different plastic material such as PVC, PMMA, PET (thick and foil substrates) [4]. The strong decrease of T in the UV range (λ < 400 nm, ITO band-band transition) protects the plastic substrates against UV radiation while that due to the carrier (λ > 900 nm) offers a sun and heat radiation protection.

The mechanical properties of the coatings deposited on PC substrates have been studied by various tests (Table 2). The adhesion is in agreement with the Tape and the lattice cut test procedure. No scratch (class 1) was observed after 10 rubbing cycles with an eraser under a load of 10 N and with a cotton cloth (25 cycles). The Taber test shows that the coatings cannot be classified as really hard materials. The hardness is 1H. Higher values are obtained when the amount of MPTS is increased, but such coatings present also a higher value of resistivity.

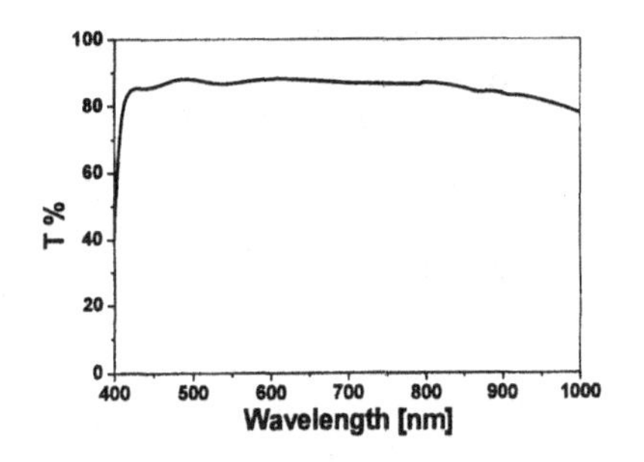

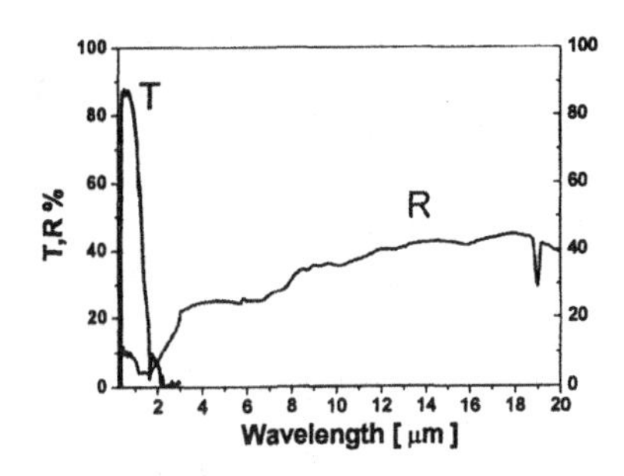

Figure 7. Visible near IR and a IR transmission and reflection spectra of a MPTS/ITO nanocomposite coating deposited on a polycarbonate substrate. The coating has been UV treated and annealed in forming gas.

Table 2. Mechanical properties of MPTS/ITO nanocomposite coatings on PC.

Adhesion		Abrasion		Hardness
a	b	c, d	e	f
OK	Gt0	Class 1	Haze 15%@10 cycles Haze 42%@1000 cycles	1 H

a) Tape Test (DIN 58196-K2), b) Lattice Cut Test (ASTM 3359, DIN 53151), c) Abrasion with cotton cloth (DIN 58196-H25), d) Abrasion with hard rubber (DIN 58196-G10), e) Taber Test (DIN 52347/CS10F/5,4N), f) Pencil Test (ASTM 3363-92a)

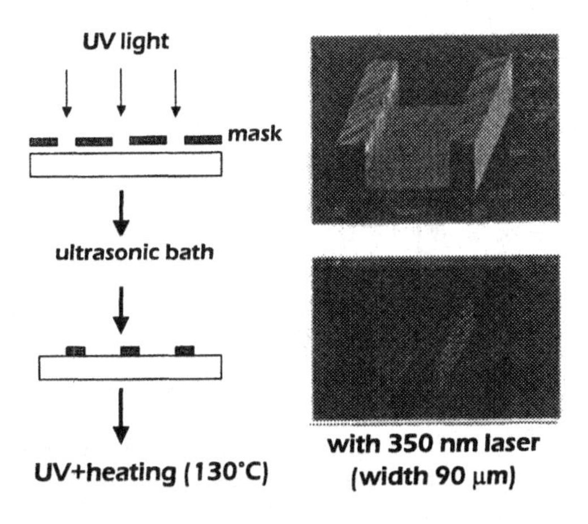

Figure 8. Line patterns obtained by selective irradiation through a mask or using a 350 nm laser, removing the non-exposed area by washing in ethanol.

The coatings are easily patterned by selective UV irradiation. The exposed parts strongly adhere to the substrate and the non-exposed parts are easily washed under ultrasound in ethanol. Figure 8 shows typical line patterns obtained by UV irradiation through a metallic mask placed directly on top of the wet coating or by a 350 nm laser irradiation.

Antiglare-antistatic obtained by a spraying process to get a final surface morphology with micrometer size roughness are reported in [3,4].

The nanotechnology concept is therefore particularly well adapted for the obtention of conducting coatings on heat sensitive substrates like plastics with stable resistivity down to about $5 \times 10^{-2}\Omega$cm, i.e. quite comparable to those obtained with conducting polymers. Moreover, their properties are stable under UV or visible light irradiation.

CONCLUSION

Very stable sols allowing the deposition of conducting, antistatic and antiglare-antistatic coatings fully processable either at low temperature ($T < 130°C$) or till high temperature (1000°C) have been developed using a nanotechnology concept. The addition of small amount of hydrolysable silanes (binders) permits the processing of any heat sensitive substrates and improves the mechanical properties of the coatings without deteriorating their optical and electrical properties. The concept can be extended to other functional coatings.

REFERENCES

1. J. Puetz, F.N. Chalvet, M.A. Aegerter, "Transparent conducting coatings on glass tubes", pp 73-80, in *Sol-Gel Optics VI*, Proceedings of SPIE, vol. 4804, Edited by E.J.A. Pope, H.K. Schmidt. B.S. Dunn, SPIE, Washington/USA, 2002.
2. J. Puetz, G. Gasparro, M.A. Aegerter, "Liquid film deposition of transparent conducting oxide coatings", pp 179-184, in *Proc. 4th International Conference on Coatings on Glass (4th ICCG)*, Edited by C.P. Klages, H.J. Gläser, M.A. Aegerter, The Organizing Committee of the 4th ICCG, Braunschweig/Germany, 2002 (also to appear in Thin Solid Films (2003)).
3. M.A. Aegerter, N. Al-Dahoudi, "Wet chemical processing of transparent and antiglare conducting ITO coatings on plastic substrates, *J. Sol-Gel Science and Technology*, 27, 81-89, 2003.
4. N. Al-Dahoudi, "Wet chemical deposition of transparent conducting coatings made of redispersable crystalline ITO nanoparticles on glass and polymer substrates", PhD Thesis, Universitaet des Saarlandes and Institut fuer Neue Materialien - INM, 2003.

MICRO STRUCTURE OF N-IMPLANTED TI THIN FILMS PREPARED BY ION BEAM SPUTTERING DEPOSITION

Shinji Muraishi
RCAST, The University of Tokyo
4-6-1 Komaba, Meguro-ku, Tokyo,
153-8904 Japan

Tatsuhiko Aizawa
CCR, The University of Tokyo
4-6-1 Komaba, Meguro-ku, Tokyo,
153-8904 Japan

ABSTRACT

Ti thin films with and without nitrogen implantation was prepared by cold-sputtering for advanced processing of nano-structured materials. Ti films were grown on (001) Si substrates by ion-beam sputter deposition. Ti target of 99.9% purity was sputtered at room temperature by Ar^+ with the beam energy of 1keV and the pressure of $5x10^{-2}$ Pa. Target and substrate were well-cooled in order to reduce the grain growth by increase of temperature. N^+ implantation was carried out for 150 nm thick Ti films with the ion-energy of 100keV and the dose of 0.6~5x 10^{17} ion/cm^2. Cross-sectional TEM observation for as-deposited Ti film revealed that columnar grains in the order of 10 nm grew with the direction perpendicular to the Si substrate. Diffraction pattern from Ti film showed the Debye-ring consistent to fcc structure, not the conventional hcp structure. After N^+ implantation, strong [111] texture with the direction parallel to [001]Si was confirmed from diffraction pattern. Formation of nitride was confirmed by annealing the higher-dose samples. Tetragonal structure of Ti_2N was detected among the tranceformed titanium phase with hcp structure. Formation of nitrides was discussed from the depth profile of nitrogen and chemical shift of Ti 2p spectrum through XPS analysis.

INTRODUCTION

Titanium nitrides exhibit superior characteristics such as extreme high hardness, high melting point, high thermal and electrical conductivity. These preferable properties were widely applied to industrial coating material for cutting tools or diffusion barrier of electric devices. TiN coating films were usually fabricated by CVD method using chemical reaction between $TiCl_4$ and NH_4[1] or by PVD method using reactive sputter in atmosphere of N_2/Ar mixture[2]. These reactive deposition techniques only produce the planar, laminated structure with combination of single phases. Ion implantation technique can make structural control for various material designs by handling the non-equilibrium phase or super saturated solid solution without loss of superior properties. Selection of chemical species and elements in the ion implantation leads to well-conditioned surface modification to protection of ceramic coating films from severe wear and oxidation[3]. In the present work, non-equilibrium Ti-N film was fabricated under cold state by combination of Ion Beam Sputter Deposition and Ion Implantation technique. Microstructure of Ti and Ti-N films and their structural changes by thermal annealing were evaluated by TEM and XPS. Formation of nitrides was mainly discussed from TEM diffraction and XPS chemical state analysis.

EXPERIMENTAL PROCEDURE

Ti thin film was prepared by Ion Beam Sputter Deposition method (IBSD). Titanium target with the purity of 99.9 % was sputtered by Ar^+ beam with the beam energy of 1keV and the beam current of 25 μA. Target and substrate were well cooled by water so that deposition was carried out under cold state. Total pressure in the working chamber was $5x10^{-2}$ Pa during deposition. Nitrogen implantation was carried out on these samples with the beam energy of 100keV and the implantation dose of 0.6 ~ 5 x 10^{17} ion/cm^2. Some of them were heat-treated for 3.6 ks at 773K in vacuum furnace with $5x10^{-4}$ Pa. Observation of microstructure of Ti films was made by TEM (JEOL 2010F). Sample preparation for TEM was performed by IV4/HL (Technoorg Linda) with 4keV Ar^+ beam at the incident angle of 5 degree. XPS measurement was carried out through Ti 2p3/2, N 1s, O 1s and Si 2p3/2 spectra by JPS 9200T(JEOL). Depth profile was measured by 3keV Ar+ beam etching. Chemical state analysis was made for Ti 2p3/2 to define the formation of nitrides in Ti films.

RESULTS AND DISCUSSION

Microstructure of Ti and Ti-N films prepared by IBSD and Ion Implantation.

Microstructure of as-deposited Ti film.

Thin Titanium film was deposited on Si substrate by IBSD to investigate the effect of cold sputter deposition on it's microstructure. Cross-sectional view of as-deposited Ti film by TEM was shown in fig.1 (a). Columnar-like grains with the average size of 10nm in diameter, grew in perpendicular to Si substrate. Faint contrast of grain boundary indicated that formation and growth of Ti crystal should be suppressed by cold state sputtering deposition. TEM diffraction pattern shows Ti has fcc structure, not the conventional hcp structure (Fig.1(b)). Measured lattice constant was a=0.438 nm. In the literature[4-5], formation of fcc Ti was reported by solution of hydrogen and nitrogen into Ti. Since nitrogen or hydrogen atoms occupy the octahedral or tetragonal sites of α-hcp Ti, the lattice constant can be modified; e.g., 0.441 nm for Ti-H solid solution. In this experiment, no oxygen and nitrogen were detected in the as-deposited Ti film by XPS measurement. Hence, this structural change might be induced by introduction of excess vacancies into thin film via the low temperature sputtering deposition.

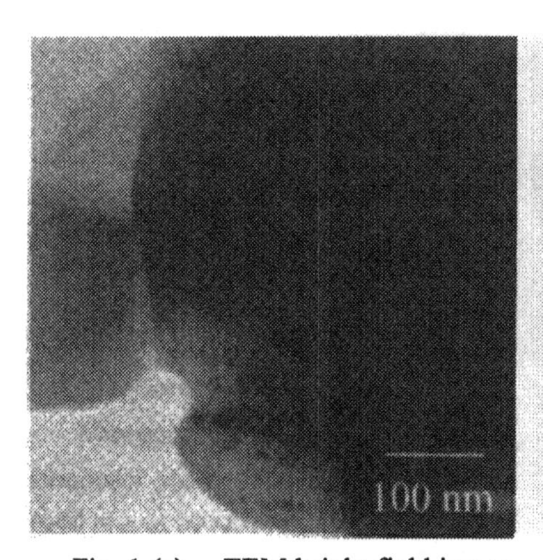

Fig. 1 (a). TEM bright field image of Ti thin film on (001) Si substrate. Columnar-grain grew perpendicular to substrate.

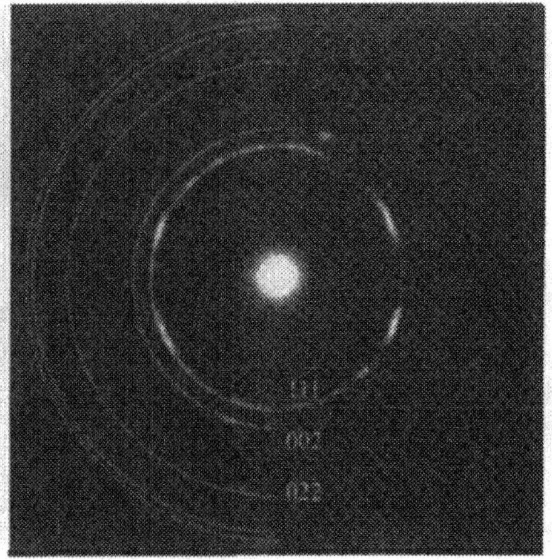

Fig. 1 (b). TEM diffraction pattern of as-deposited Ti film showed fcc structure.

Microstructure of N-implanted Ti film with the low implantation dose.

Nitrogen implantation was carried out for fcc Ti film to investigate the structural change due to the nitrogen implantation. Figure 2 showed TEM dark field image of N-implanted Ti film with the dose of 6.0 x 10^{16} ion/cm^2. Nitrogen implantation was made with the ion energy of 100 keV. Since the edge portion of the film was thinned down during TEM sample preparation, the film thickness was reduced to 80 nm. Amorphous layer of 100 nm in thick was observed in Si region. From the SRIM simulation, the penetration depth of nitrogen was calculated to be 210 nm in maximum, 150nm in average. Hence, nitrogen reached to affect Si crystal in the present condition. In Ti region, fine grains of 10 nm were clearly observed as bright contrast. Diffraction pattern showed N-implanted Ti has also fcc structure with strong {111} texture parallel to the substrate. Lattice constant was measured to be a= 0.418 nm. This value was smaller than that of as-deposited Ti film (a=0.438). In the literature[4], the lattice constant of fcc Ti-N solid solution was reported to be a=0.423 nm for nitrogen implanted epitaxial Ti film with the dose of 6x10^{17} N/cm^2. Relatively small lattice constant was observed in the present study even by the lower dose of nitrogen implantation. Nitrogen might be effectively implanted into fcc Ti film to form solid state in fcc Ti film.

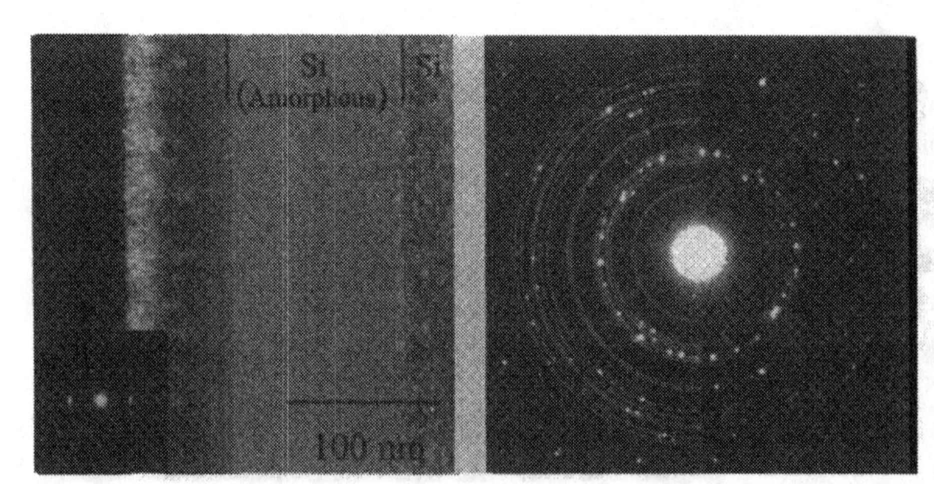

Figure 2. TEM dark field image of N-implanted Ti with dose of 5x10^{16}. Si substrate was damaged and form amorphous layer by ion implantation.

Figure 3. Diffraction pattern of annealed N-implanted Ti film with 5x10^{17} dose. Tetragonal structure of Ti_2N was recognized.

This is also supported by the fact that nitride formation is hardly confirmed from diffraction pattern. Hence, titanium in fcc structure might be favored for solid solution formation of fcc Ti-N with higher nitrogen content.

Microstructure of N-implanted Ti film with high implantation dose.

In order to promote the nitride formation, the implanted nitrogen dose was increased to $5x10^{17}$ ion/cm^2 and additional heat treatment was made. The dose of $5x10^{17}$ ion/cm^2 is corresponded to Ti-40at%N alloy in average. In literature[6], significant reaction between Ti and Si was reported above 923 K. Taking into account for the Ti/Si reaction across the interface, Ti-N film was annealed at 773 K for 3.6 ks. The solubility of nitrogen in Ti was 5at% at 773 K. Figure 3 showed diffraction pattern obtained from Ti-N film aged at 773K for 3.6 ks. Tetragonal symmetry was clearly observed among hcp Ti. This tetragonal structure is just a proof of Ti_2N formation (a=4.9452 c=3.0342). Since the solubility of nitrogen in Ti was only 5a% at 773 K in the phase diagram, the solid solution with high nitrogen content might be modified by this annealing. Hence, non-equilibrium cubic structure of Ti-N solid solution was decomposed to Ti_2N and hcp Ti by the subsequent heat treatment to implantation.

XPS measurement of Ti-N film.

Nitrogen depth profiling through XPS measurement.

XPS measurements were performed for the implanted samples with the nitrogen dose of $1x10^{17}$ and $5x10^{17}$ ion/cm^2. Nitrogen concentration in depth was measured by XPS as shown in Fig. 4. Four spectra of Ti 2_P, N 1_S, O 1_S and Si 2_P were measured after Ar beam etching with the energy of 3keV. Judging from survey counts of Si spectrum, the sputtering rate of Ti was estimated to be about 0.01nm/sec. Ti/Si interface corresponded to $1.5x10^4$ sec. Oxygen was only detected at the film surface and it existed as oxides or adsorbed molecules. Nitrogen concentration has maximum in the vicinity of Ti/Si interface. The nitrogen depth profile by XPS measurement was also in agreement with SRIM simulation results. Figure 5 showed the depth profile of nitrogen concentration in Ti for two doses. Total nitrogen penetrated in Ti film was calculated to be Ti-20%N for $5x10^{17}$ N/cm^2 and Ti-6%N for $1x10^{17}$ N/cm^2. These concentrations were about the half of theoretical values estimated from the nitrogen doses; i.e., Ti-40%N can be obtained by $5x10^{17}$ N/cm^2 and Ti-10%N, by $1x10^{17}$ N/cm^2. Since

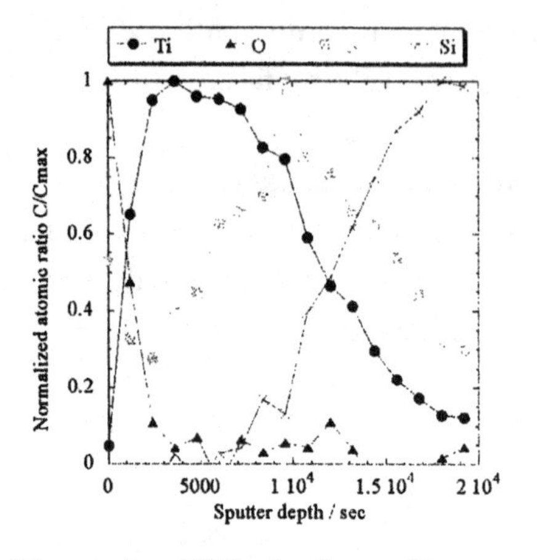

Figure 4. XPS depth profile was carried out for the sample of 5x10^{17} dose. Atomic ratio was normalized by it' maximum value.

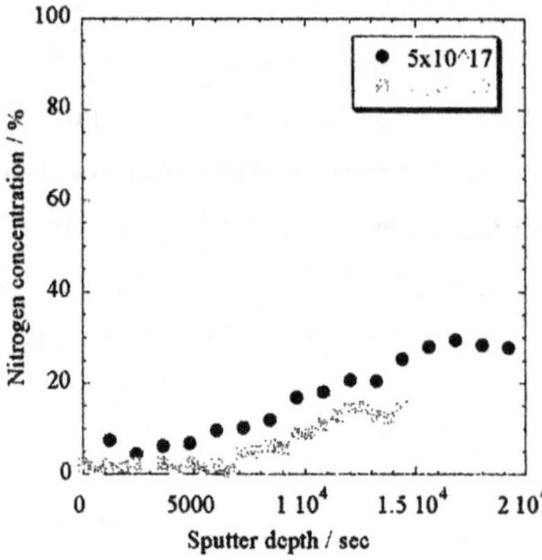

Figure 5. N vs. Ti concentration was calculated for different N-dosed samples. 30% of N in Ti was achieved at maximum point for 5x10^{17} dose.

the maximum penetration depth crossed over Ti/Si interface and reached to deeper region in Si, about a half amount of nitrogen atoms were possibly trapped in Si region. Maximum nitrogen concentration in Fig.5 was attained to be Ti-30at%N for 5x10^{17} N/cm^2, Ti-15at%N for 1x10^{17} N/cm^2. In the literature[7], formation of nitrides through nitrogen implantation was reported for bulk titanium with the purity of 99.5%. They reported that nitride formation was dependent on the implantation dose; i.e., inflection point was observed at 30at%N in the plotted relation of the binding energy difference between Ti 2p3/2 and N 1s peak shifts against the nitrogen content. This inflection point corresponded to a branched better TiN and Ti_2N formation processes. Since the present condition was set-up in the intermediate region between Ti_2N and TiN formation, Ti_2N compound might be selectively formed in particular for the implanted sample with the dose of 5x10^{17} ion/cm^2.

Chemical state analysis for Ti2p3/2 spectra related with Ti-N bonding.

Chemical state analysis was made for higher dosed samples with and without heat treatment. Heat treatment was made at 773K for 3.6 ks. Peak shifts of Ti 2p3/2 spectrum were plotted against the sputtered depth in Fig. 6. In the present study, binding energy of Ti-Ti bonding was observed at 453.9 eV. Large chemical shift was seen at the beginning of sputtered depth (3000 sec); i.e., surface oxides were formed. After 3000sec or except for the oxidation affected layer, the binding energy gradually increase with the sputter depth. This increase of the chemical shift was in accordance with the increasing nitrogen concentration in Fig. 5 and 6. The binding energy of annealed samples was slightly higher than that of as-implanted samples; maximum increase of binding energy (ΔBE) was 0.3eV for as-implanted samples and ΔBE = 0.35eV for the annealed ones. This might be higher nitrogen concentration enhances nitride formation for the solid solution. Since diffusion of nitrogen was easily occurred at higher temperature, diffusion might induce the nitride formation for annealed sample. After 15000 sec, binding energy of the annealed sample was decreased in the vicinity of interface. In this region, peak shift of Si 2p spectrum was observed at 101 eV. In the literature[8], binding energy shift due to Si-N bonding was reported for Si 2p spectrum at TiN/Si interface. Ti-Si bonding was induced by diffusion at the interface region, Si might work as a reductive element against the Ti-N bonding. In the literature[7], nitride formation was well related to the binding energy shift; e.g., formation of Ti_2N was certified by the binding energy shift of 0.4 eV for bulk Ti with the dose of $5x10^{17}$ ion/cm^2. Even from the binding energy shift, formation of Ti_2N was recognized after annealing.

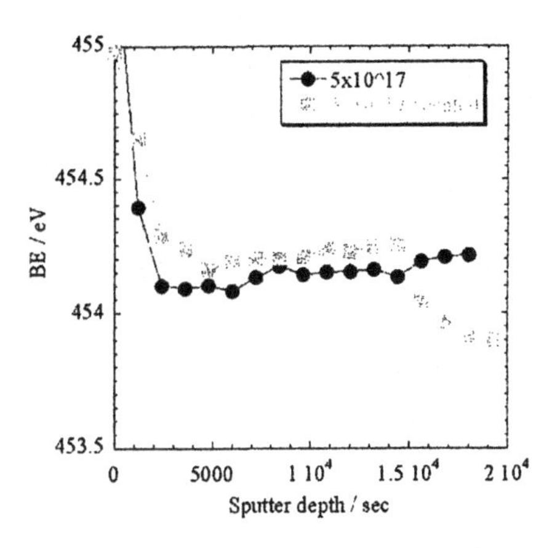

Figure 6. Ti 2p3/2 chemical shift was measured for the sample with $5x10^{17}$ dose with and without heat treatment.

CONCLUSION

Structural controlling of Ti and Ti-N film was investigated under cold state by IBSD and Ion Implantation technique. TEM observation revealed that as-deposited Ti film showed fcc structure, not conventional hcp. Non-equilibrium fcc structure might be induced by introduction of excess vacancies into Ti film under cold sputtering deposition. By implantation of nitrogen ions into fcc Ti film, cubic structure of Ti-N solid solution was obtained with strong {111} texture. Non-equilibrium structure of Ti-N with the dose of $5x10^{17}$ N/cm^2 decomposed to tetragonal Ti_2N and hcp Ti-N by subsequent heat treatment. From the XPS measurement for Ti-N film with and without heat treatment, Ti 2p3/2 binding energy shift of annealed Ti-N was slightly higher than that of as-implanted Ti-N. Chemical shift due to Ti-N bonding assumed the formation of Ti_2N.

REFERENCES

[1] K. Hiramatsu, H. Ohnishi, T. Takahama and K. Yamanishi, "Formation of TiN films with low Cl concentration by pulsed plasma chemical vapor deposition", Journal of Vacuum Science Technology A **14** [3] 1037-1040 (1995).

[2] E. Galvanetto, F. P. Galliano, F. Borgioli, U. Bardi and A. Lavachi, "XRD and XPS Study on Reactive Plasma Sprayed Titanium-Titanium Nitride Coatings", *Thin Solid Films* **384** 223-229 (2002).

[3] A. Mitsuo and T. Aizawa, "Effect of Chlorine Distribution Profiles on Tribological Properties for Chlorine-implanted Titanium nitride films", *Surface and Coating Technology* **158-159** 694-698 (2002).

[4] Y. Kasukabe, A. Ito, S. Nagata, M. Kishimoto, Y. Fujino, S. Yamaguchi and Y. Yamada, "Control of Epitaxial Orientation of TiN Thin Films Grown by N-implantation", *Applied Surface Science* **130-132** 643-650 (1998).

[5] P. Scardi, M. Leoni and R. Checchetto, "Residual Strain in Deuterated Ti thin Films", Matterials Letters **36** 1-6 (1998)

[6] T. Stark, L. Gutowski, M. Herden, H. Grunleitner, S. Kohler, M. Hundhausen and L. Ley, "Ti-silicide formation during isochronal annealing followed by in situ ellipsometry", *Microelectronic Engineering* **55** 101-107 (2001).

[7] K. Terada, K. Matsushita and T. Minegishi, "XPS Analysis of Titanium Nitride and Zirconium Nitride Compound Thin Layer Formed by Nitrogen Ion Implantation", *The Journal of The Surface Finishing Society of Japan* **43** [1] 19-23 (1992).

[8] P. Y. Jouan, M. C. Peignon, Ch. Cardinaud and G. Lemperiere, "Characterisation of TiN Coatings and of the Ti/Si Interface by X-ray Photoelectron Spectroscopy and Auger Electron Spectroscopy", *Applied Surface Science* **68** 595-603 (1993)

NOVEL NANOTECHNOLOGY OF USABLE SUPERCONDUCTOR CERAMICS

Anatoly E. Rokhvarger* and Lubov A. Chigirinsky
Polytechnic University, 6 Metro Tech Center, Brooklyn, NY, 11201, USA

ABSTRACT

A novel ceramic-silicone processing method brings superconductor properties of $YBa_2Cu_3O_{7-x}$ ceramic raw powder to nano-textured and sintered micro-grains of adhesion coated substrate, slip/tape cast and bulk articles. This paper presents an understanding of the efficiency of the applied thermal, chemical and physical methods of the phase transition and 3D nano-crystal structure evolution that cause extraordinary quality of superconductor macro-material and products.

INTRODUCTION

Over six thousand years of ceramic technology methods are used to convert metal oxide powder particles into artificial stone-like products of required shapes and properties. Recently developed electrical and electronic ceramic technologies used in the production of semiconductors and piezoelectric, capacitor and insulation ceramics are continuing this traditional ceramic engineering method by bringing electricity-related properties of the particular ceramic crystals to the sintered crystal populations of the ceramic engineering materials and products. However, until now no one could achieve similar practical results using high temperature superconductor (HTS) ceramic powder and particularly modified traditional methods of ceramic technology, probably because of the complexities of superconductor physical theories and specific structural requirements.

This paper continues a cycle of our patents, presentations and publications [1–5] reflecting a successful finish of the multi-disciplinary feasibility research to develop the cost-effective nanotechnology of inexpensive HTS nano-structured composite material and highly marketable electrical and electronics products from this material using readily available $YBa_2Cu_3O_{7-x}$ (YBCO) fine powder [6].

RESULTS OF THE RESEARH

The invented HTS ceramic-silicone processing (CSP) we named HTS-CSP method. It results in HTS-CSP composite material and usable products and consists of six innovative stages:

(i) preparation and ultrasonic homogenization of nano-size colloid slurry of YBCO and silver dope fine powders in a silicone-toluene solution;
(ii) versatile material forming using the following methods: a) adhesion substrate coating, such as dip coating, brushing, ink printing, painting, and pulverizing; b) slip and tape casting in plastic forms; c) extrusion or injection molding of the plastic mass from condensed slurry; and d) pressing of the dried plastic mass;
(iii) magnetic crystal and grain orientation of green coated and cast materials;
(iv) thermal polymerization to consolidate YBCO crystals and organize them in a superconductive effective crystal texture;
(v) full dense incongruent sintering by silicate glass eutectics that are products of thermo-chemical reactions of silicone burnt residuals with metal oxides of YBCO;
(vi) thermal oxygenation to cure YBCO orthorhombic crystal morphology.

Two specific features determined the effectiveness of the HTS-CSP method. The first one was a choice for further use of fine YBCO powder following ultrasonic dispersion of YBCO particles and crystal conglomerates in silicone-toluene solution to get a nano-size suspension slurry. Only nano-sizes of YBCO grains allowed achieving and completing their magnetic orientation and homogeneous ceramics filled polymer organization as well as full dense eutectic sintering and complete oxygenation of the YBCO crystals. Indeed, our achieved positive results strictly depended on a vast surface value of the used ceramic particles and consequently their reactivity during technologically induced magnetic and other thermo-chemical and oxygenation impacts.

The second specific feature of the HTS-CSP method is a newly discovered multifunctional silicone polymer additive [1, 2], which plays at least twelve functional roles [3]. This component of HTS-CSP material formulation determines workability and allows chemically self-organized process and quality control of the developed nanotechnology. It is an inexpensive liquid silicone polymer $HO\text{-}[\text{-}Si(CH_3)_2O\text{-}]\text{-}H$. Silicone does not react with and prevents degradation of HTS ceramics during initial technological treatment. Silicone burnt residuals produce silicate glass eutectics hardening as thin nano-films and points within the crystal and grain boundary areas. These increase durability and improve the mechanical characteristics of the novel HTS material. We apply silicone in an amount of 1% – 10% of the YBCO ceramics weight. Additionally, we use polymerization additives and some dopes, for example, silver powder.

The resulting colloid slurry was used to provide the dip adhesion coating or spraying or ink printing of metal alloy, ceramic or quartz glass substrates of any size and form including silver and nichrome strands for multi-filament round wire [2, 3]. The same colloid slurry may be used for slip and tape casting of flat or curved articles. If HTS-CSP slurry is condensed into a ceramic mass, it is suitable for extrusion or dry pressing of bulk articles, such as beams, rods, tubes and disks. Additionally, HTS-CSP slurry can be used in electronics for multi-layer ink printing, brushing, painting and lithographic techniques.

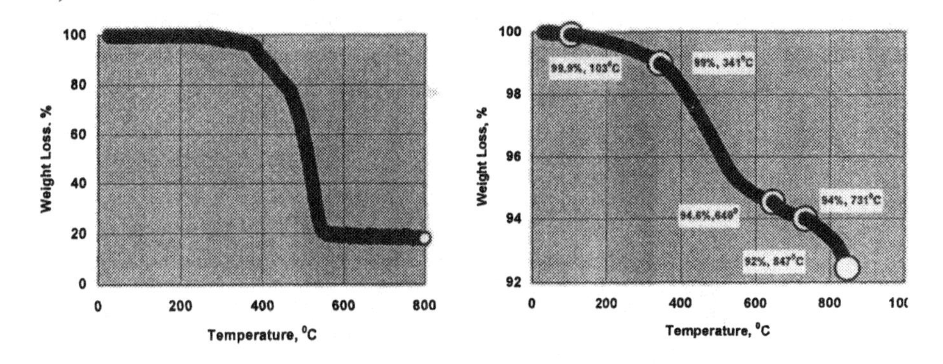

Figure 1. TGA in airflow of silicone resin with a cross-linker (left).
Figure 2. TGA in nitrogen flow of the green (before polymerization) HTS-CSP composite mass (right).

As shown in Figure 1 of the thermal gravimetric analysis (TGA), silicone polymer mass includes ~20% inorganic silicon (Si) atoms that have to be homogeneously distributed within the sintered ceramic body. 80% of the silicone is the burnt out organics. If our raw material composition comprises of 5% silicone additive, the organic part including toluene solvent and other volatiles decrease the weight of the green HTS-CSP composite on 5.4% when it is heated up to 640^0C. Further heating induces destruction of the YBCO crystals and loss of a particular part of the oxygen atoms (Figure 2).

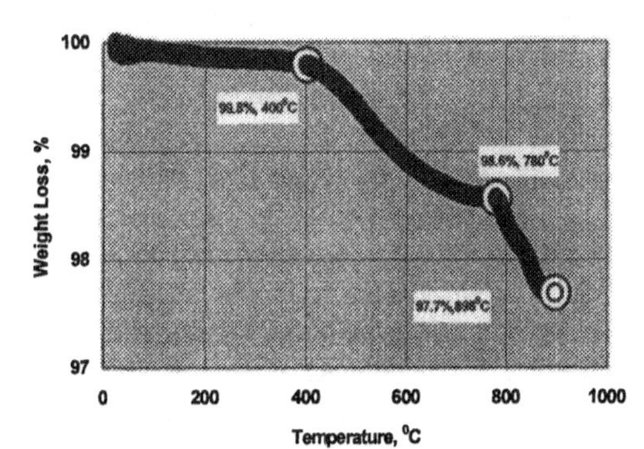

Figure 3. TGA in nitrogen flow of HTS-CSP sintered composite.

We used TGA data to develop a schedule of the thermal processing of HTS-CSP composite including cooling of the sintered HTS-CSP material. This stage was accompanied with a thermal oxygenation impact to rebuild superconductive morphology of the YBCO crystals.

As shown in Figure 3, sintered and oxygenated HTS-CSP composite loses ~ 2.2% of the total weight, which determines an amount of the thermodynamically reversible oxygen responsible for rebuilding orthorhombic crystal morphology and achieved superconductivity of the HTS-CSP composite material.

Our publication [4] focused on the invented method of the nano-particles magnetic orientation and 3D cross-linking polymer induced consolidation and organization of the YBCO crystals resulting in nano-texturing of the HTS-CSP sintered ceramic body. The AFM image in Figure 4 shows the effectiveness of the applied magnetic-polymer *c*-axis crystal orientation and organization method.

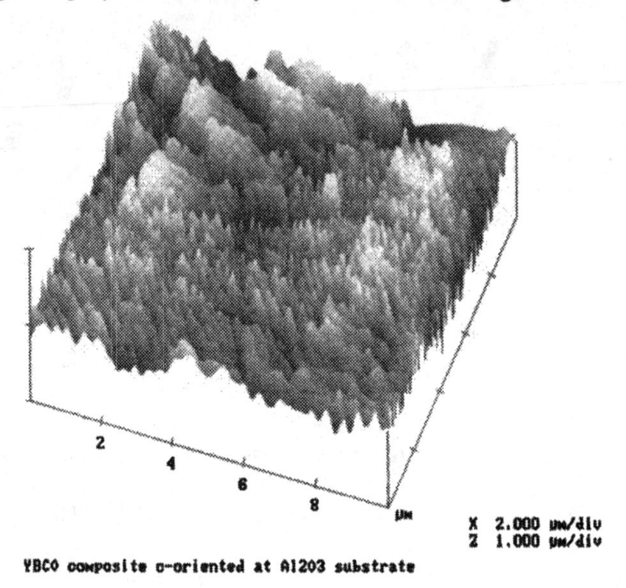

Figure 4. AFM surface image of the sintered HTS-CSP sample: adhesion coating/painting on ceramic substrate.

After thermal treatment shrinkage, the self-controlled thickness of the HTS-CSP adhesion layers was about 10μm [3]. HTS-CSP composite/substrate cross-section ratio for nichrome substrate strand of 50μm in diameter is about 1:1, which is highly beneficial from the point of view of the HTS-CSP multi-strand round wire. As shown by the SEM, the cross-section of the HTS-CSP coating ring consists of three parts: (i) bonding interlayer of substrate material with HTS-CSP composite; (ii) central full-dense HTS-CSP layer, and (iii) outside rough surface layer. These provide a layer thickness ratio 1:2:1. Since the bonding layer is probably not superconductive and considering the roughness of the outside layer, the superconductor working thickness should comprise 50% - 70% of the total thickness of the sintered HTS-CSP coating layer. Additionally, there are the silicate glass hardened eutectics in forms of nano-films and dots and silver dope that are necessary for HTS-CSP material processing and use, totally about 3% - 7% of HTS-CSP composite weight or 5% - 10% of the volume. These as well as sintering uncertainties and "broken" YBCO crystals comprise some unknown "non-working" part of the HTS-CSP composite intermixture and should attenuate the superconductive behavior of the HTS-CSP composite.

To check the significance of such "ballast" and its impact on HTS-CSP material superconductivity, a few different HTS-CSP material samples have been verified at Brookhaven National Laboratories, NY, USA, (courtesy of Dr-s M.

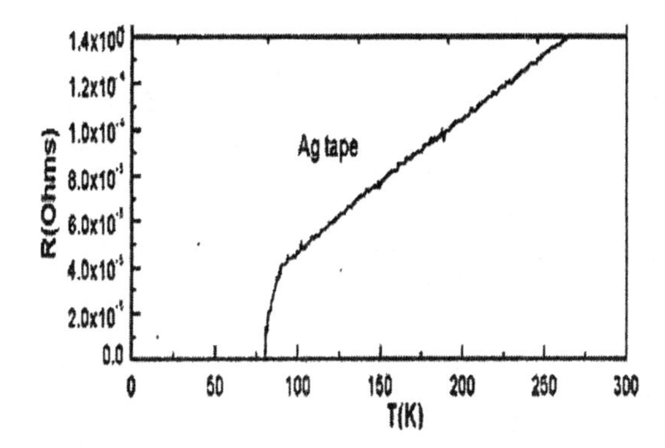

Suenaga and S. Solovjov, Figure 5) and at the Russian Academy of Science, Moscow (courtesy of Dr. A. Ionov, Figures 6 and 7). As show Figures 5–7, superconductive behavior of the HTS-CSP material is similar to $YBa_2Cu_3O_7$ crystal behavior with just a few differences.

Figure 5. The electrical current resistance of the HTS-CSP adhesion coated tape on silver substrate as a function of temperature.

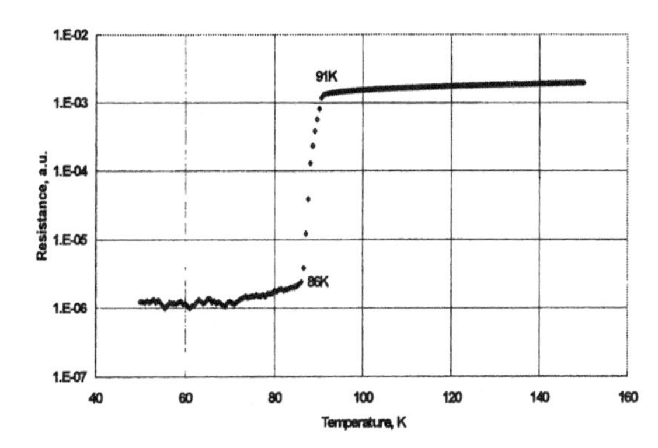

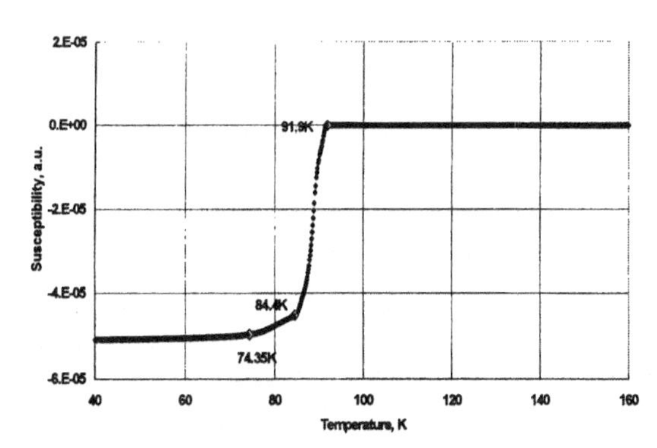

The reasonable engineering question is whether and how these particular differences influence the workable characteristics and practical use of HTS-CSP macro-material in liquid nitrogen (LN) coolant ambience. To provide a quantifiable answer to the above question, we measured electric current carrying capacity (J A/cm^2) of the HTS-CSP samples [2]. For comparison purposes all measurements of the HTS-CSP samples including adhesion coated and initial (non-coated) silver and nichrome strands, slip/tape cast strips and pressed tablets were provided at both room and LN temperatures.

Additionally, slip cast samples showed good Meissner effect.

Figure 6. Resistance in arbitrary units vs. temperature for the HTS-CSP slip cast
Figure 7. Magnetic susceptibility in arbitrary units vs. temperature for the similar slip cast sample.

As shown by X-ray analysis [4] and evidenced by superconductor behavior of the HTS-CSP samples reflected in Figures 5 – 7, sintered macro-bodies of HTS-CSP samples mostly comprised of nano-crystal grains/conglomerates with prevailing contents of $YBa_2Cu_3O_7$ crystals and their portions were big enough to realize percolation effect and overcome a sensitivity threshold of the measurement systems being used. Meanwhile, HTS-CSP macro-composite demonstrates superconductor temperature transition width larger than for $YBa_2Cu_3O_7$ crystals.

We used the four-point engineering method of ASTM B714-82 (90) to define Volt (***E***) – Ampere (***I***) characteristics (***E*** *vs.* ***I***) of the HTS-CSP samples in dependence on applied ***I***, where ***E*** is an electric field measured with precision 0.1μV/cm and ***I*** is an electric transport current, which we increased up to 20 A. ASTM based measurements of the electric current transferring through *macro*-body of the studied material uses to use samples with current lead length > 1.5cm.

Using conventional design parameters of electric and electronic systems that are now working with copper conductors at room temperature and at the particular ***W*** (heat dissipation), we compared ***E–I*** data of ordinary copper wire and HTS-CSP samples tested at LN temperature. For example, in our ASTM measurements of HTS-CSP adhesion coated silver strand of 0.127mm in diameter and sintered layer thickness 10μm transferred electric current of 18A at ***E*** = 0.02V/cm providing energy losses lower than engineers mean acceptable for some electrical engineering applications, such as ***W*** = 10 Watt/cm^3. The estimated electric current carrying capacity of the total cross-section of the measured HTS-CSP strand was equal to ***J*** = 108kA/cm^2 at 77K. With the same ***W*** and ***E***, the HTS-CSP adhesion coated nichrome strand of 50μm in diameter and the same coating layer thickness of 10μm transferred electric current of 2A, which corresponds with ***J*** = 52kA/cm^2. Meanwhile copper wire can transfer at these ***W*** and ***E*** only 500A/cm^2. Therefore it can be seen that a copper conductor has correspondingly less carrying capacity than these two HTS-CSP filament samples by 216 and 104 times.

These are well in excess of requirements for beneficial application of HTS-CSP multi-strand wire in electrical industry applications thereby reducing cost, weight and size of electrical motors, generator rotors, transformers, and cables by at least 10 - 15 times. If some electronics need, for example, insignificant heat dissipation at ***W*** = 0.1Watt/cm^3 and ***E*** = 0.5mV/cm, ***J*** of the same HTS-CSP adhesion coated filaments on silver and nichrome substrate strands exceeds ***J*** of the corresponding copper or silver leads by 2.5 and 11 times. These can cause miniaturization of electronics of 1.5x – 3.0x, which is also very beneficial.

DISCUSSION

A subscripting stoichiometric coefficient at the *O* atom in the formula $YBa_2Cu_3O_{7-x}$ of the YBCO superconductor ceramics is a statistical estimation of the mean value of the oxygen content in the particularly studied *micro* grain/conglomerate crystal combination comprising of the crystals with different morphology [7]. Only $YBa_2Cu_3O_7$ orthorhombic crystals have critical temperature of the electric current transition in superconductive state at T_c = 93K. In the range

$-0.2 \leq x \leq 0.2$, an impact of the particular part of the $YBa_2Cu_3O_7$ crystals on the crystal *micro*-conglomerate is enough to demonstrate the engineering acceptable superconductivity at 77K since 77K is the boiling temperature of the liquid nitrogen (LN), which is an inexpensive coolant. Meanwhile, electric current density can practically vary within the range of $10^6 A/cm^2 \geq J \geq 0 A/cm^2$ depending on several characteristics of crystal conglomerates including grain boundaries, dopes and additives, grain homogeneity, and crystal-grain topology.

Provided ASTM measurements integrate electrical current fluxes. For HTS-CSP wire these fluxes are automatically distributed between a conducting metal core substrate and various superconductor channels of the coating layer in correspondence with their resistances in bottleneck cross-sections. Places of bottleneck cross-sections are changeable and depend on applied I while all channels work in parallel. Therefore, the measured ***E–I*** characteristics have to be in agreement with Kirchoff's first law causing some crooks on ***E–I*** graphs.

The x variation in $YBa_2Cu_3O_{7-x}$ conglomerates in synergy with uncertainties in grain boundaries and incorporation of impurities causing by thermo-chemical reactions of technological additives and dopes with YBCO crystals, unavoidably attenuate macro-material superconductivity. As a result of such attenuation, second order polynomial equations approximating ***E–I*** graphs of the HTS-CSP macro-samples have less coefficient values at the second order polynomial terms than the same for ***E–I*** graphs of $YBa_2Cu_3O_7$ crystals.

Fortunately, ASTM measured engineering attenuation of superconductivity of sintered substrate coated strands, slip/tape cast and other bulk HTS-CSP macro-products is much less of the threshold for high beneficial engineering applications and sure competition of HTS-CSP products with copper leads. For example, whole cross-section of the HTS-CSP adhesion coated silver strand of 0.127mm in diameter and sintered layer thickness 10μm transfers J =108kA/cm^2 at W = 10Watt/cm^3. At such conditions the middle "mostly working" part of the HTS-CSP coating layer comprises ~10% whole cross-section of the HTS-CSP coated strand and can transfers at least 90% the total electric current or $J \sim 9 \times 10^5 A/cm^2$. Consequently, a copper conductor has 1800x less electric current carrying capacity than this "mostly working" part of the HTS-CSP composite material .

While $YBa_2Cu_3O_{7-x}$ ceramics are type II superconductors, HTS-CSP composite in form of slip cast and other bulk samples demonstrates significant Meissner (magnetic levitation) effect probably due to silicate glass film and silver dope nano-impurities at grain boundary areas that can provide inter-granular or "Josephson" magnetic flux pinning centers.

CONCLUSION

1. Our novel material formulation uses, as a major component, market available YBCO ceramic fine powder, a multi-functional silicone polymer additive, and some dopes. This formulation makes possible technologically guided thermo-chemical phase transition, crystal morphology transformation and 3D self-assembling nano-structuring evolution of YBCO *c*-axis oriented crystal

grain topology. HTS-CSP material is sintered in grain boundary areas by nano-thick films and dots of silicate glasses.

2. The invented HTS-CSP nanotechnology reliably brings, with just small attenuation superconductive properties of the nano-size $YBa_2Cu_3O_7$ ceramic crystals, to micro-scale crystal grains/conglomerates building a macro-scale HTS-CSP composite material usable for electrical and electronic engineering products.

3. The novel HTS-CSP nanotechnology makes possible cost-effective production of HTS-CSP ceramics in multifold forms and shapes with required mechanical properties and durability of the superconductive materials and products for all traditional and advanced electrical and electronic end user needs. Developed HTS-CSP adhesion coated round wire is flexible, inexpensive and demonstrates 500x electric current throughput in comparison with copper wire. Achieved consumer properties make HTS-CSP material and products from it highly marketable in comparison with traditional copper and silver leads and any developing technique for electric lead products and electronics.

ACKNOWLEDGEMENTS

The authors thank for their assistance Prof-s K. Levon, M. Rafailovich, A. Ulman and E. Wolf, Dr.-s A. Ionov, S. Soloviev, M. Suenaga, and Mr. A. Goldberg.

REFERENCES

1. M.I. Topchiashvili and A.E. Rokhvarger, US Patents 6,010,983 (Jan. 4, 2000) and US Patent 6,239,079 (May 29, 2001)

2. A. Rokhvarger, L. Chigirinsky and M. Topchiashvili, 2001, "Inexpensive Technology of Continuous HTS Round Wire", *The American Ceramic Society Bulletin*, **80**, No.12, pp. 37 - 42; and the same in www.ceramicbulletin.org

3. A. Rokhvarger and L. Chigirinsky (2003)"Adhesive Coated HTS Wire and Other Innovative Materials", *Proceedings of High Temperature Superconductors,* Ed. by A. Goyal, W. Wong-Ng, M. Murakami, and J. Driscoll, Vol. **140**, Am. Ceramic Soc., Westerville, OH, pp. 375 – 384.

4. A. Rokhvarger and L. Chigirinsky (2003) "Unconventional Nanotechnology of Superconductor Ceramic Articles", *Materials Research Society (MRS)* Spring Meeting, Symposium Q, April 21 – 25, 2003, San Francisco, CA – *in print and electronic format* http://www.mrs.org/publications/epubs/

5. A. Rokhvarger and L. Chigirinsky, "Cost Effective Technology of HTS Ceramic Filaments and Other Materials", An International Conference on Advanced Ceramics and Glasses, Am. Ceramic Soc., PAC RIM 4, Section 4. High T_C Superconductors – Novel Processing and Applications in the New Millennium, Nov. 4-8, 2001, Maui, Hawaii, Abstract Book, p.66

6. "*10th Anniversary Edition Product Guide,*" Superconductivity Components, Inc., Columbus, Ohio, 79 pages.

7. http://www.ill.fr/dif/3D-crystals/superconductors.html.

Industrial Development and Applications of Nanomaterials

CERAMIC NANOPARTICLE TECHNOLOGIES FOR CERAMICS AND COMPOSITES

Helmut Schmidt, Frank Tabellion, Karl-Peter Schmitt and Peter-William Oliveira
Institute for New Materials
Im Stadtwald, Geb. 43 A
66123 Saarbrücken
Germany

ABSTRACT

Nanoparticles provide interesting processing routes and material properties for many composites. For the fabrication of ZrO_2 nanoparticles, a precipitation process followed by a hydrothermal treatment has been developed. For surface modification and deagglomeration to the primary particle size, a ball milling process and a kneading process have been developed, leading to agglomerate-free particle systems between 6 and 10 nm. For photo induced diffusion, double bond containing modifiers have been used and a process for holographic pattern formation has been developed. The ZrO_2 nanoparticle systems have been used in compounding technologies for polycarbonate, showing a strong increase of the compressive strength and for the formation of ceramic membranes by a tape casting and lamination technology.

INTRODUCTION

The fabrication of nanoparticles has become of very broad spread interest for scientists and also for industry. Many types of particles are offered on the market, for example, derived by spray pyrolysis (Degussa) or similar techniques (Nanophase Inc. Ltd.). The fabrication of nanoparticles by other routes like chemical precipitation, laser pyrolysation, plasma techniques or flame spray techniques is laid down in numerous papers which, for example, are represented to a great deal in the powder conferences [1, 2, 3], and literature cited herein [4, 5, 6, 7]. However, the utilization in industrial applications of nanoparticulate system is still on a very low level. One of the main reasons for this is the fact that material costs are still very high and second, processing techniques suitable for the end user, that means the manufacturer of the final products are not available or not yet developed. This includes the whole route from the particle fabrication, the particle handling, the particle processing to materials, components and systems. One of the difficulties is related to the particle to particle interaction by van der Waals and chemical forces which are a function of the particle diameter, the increase of which is a quadratic function of the particle diameter.

For this reason, it seems to be of interest to use colloidal chemical routes including surface modification in order to tailor the particle surface chemistry for appropriate processing. Basic considerations already have been published in [8, 9, 10, 11]. In these papers, it could be shown that nanoparticle fabrication by precipitation in analogy to the sol-gel process [12] advantageously can be used to fabricate nanoparticles with controlled surface chemistry. In this paper, two alternative routes are described for the fabrication of nanoparticles also on a large scale, and some examples of the utilization of such nanoparticles using very specific surface modification is given.

EXPERIMENTAL

Nanoparticle fabrication

For powder preparation, a precipitation process under controlled particle growth condition has been used as described elsewhere [13]. A solution of zirconium n-propoxide in ethanole was added drop wise to an aqueous ammonia solution (pH > 12) containing 10 wt.% of the surface modifying agent, bifunctional amines, β-diketones or amino carbonic acids (e.g. β-Alanine) with respect to the oxide. The weight ratio of precursor and water phase was 1 : 1.

Crystallisation of the nanoparticles

The so prepared suspensions were treated at 250 °C and 80 bar for 30 minutes in a continuous working flow reactor. The resulting powder was washed to remove the no longer required processing additives and freeze-dried.

Two routes have been used for the surface modification and deagglomeration:

a) One kilogram of the freeze dried powder, organic acids (10-15 wt.% with respect to the oxide) as surface modifiers and 80 ml water were mixed and kneaded in a water cooled laboratory double-Z-kneader (Linden LK II 1) for 4 hours. During the kneading process agglomerates in the high viscous paste were deagglomerated due the high shear forces of the rotor blades. A reagglomeration is prevented through the presence of the very reactive surface modifier, which is passivating the newly generated particle surface. After the kneading process water was added to get a fully dispersed suspension with low viscosity; this suspension was freeze dried. Particle size distributions of the resultant suspensions were determined by dynamic laser light scattering (UPA).

b) The chemo mechanical comminution reactions were carried out in a water cooled Drais Perl Mill PML-H/V equipped with a zirconia milling chamber (volume: 1 litre) and a zirconia rotor. 1700 g zircon milling balls (diameter 0.3-0.4 mm) were utilized for the experiments. Distilled water, surface modifier (formic acid [5 wt.%], acetic acid [10 wt.%], 3-oxabutanic acid [10 wt.%], 3,6-dioxaheptanic acid [10 wt.%], 3,6,9-trioxadecanic acid [15 wt.%], N-(2-hydroxyethyl)ethylendiamine-N,N',N'-triacetic acid (HEDTA) [15 wt.%], N-(2-hydroxyethyl)-imino-diacetic acid (ethanoldiglycine) [15 wt.%], N,N-bis(2-hydroxy-ethyl)glycine (bicine) [15 wt.%] and 6-aminohexanic acid [15 wt.%]) and the hydrothermal produced zirconia (BET 150 m^2/g) were premixed and mechanical stirred for 30 min. Values in brackets are wt.% surface modifier used with respect to the oxide. These suspensions were milled for 4 h. Particle size distributions of the resultant suspensions were determined by dynamic laser light scattering (UPA) [14].

ZrO_2 holographic sol preparation

Methacryloxypropyl trimethoxysilane (MPTS, I) was used as matrix material in combination with zirconium propoxide (II) complexed with methacrylic acid (MA, III). In the first step, 24.80 g of MPTS is hydrolysed and condensed by a slow addition of 2,70 g of 0.5 N HCL. The water content is monitored by Karl-Fischer titration to determine the time at which water content reaches its minimum. In a second step, 4,57g of zirconium propoxide is complexed with 1.72 g of III. The complexed Zr alkoxide was mixed with the partially condensed silane under stirring conditions, and 0.54 g water (50 % of the amount necessary for stoichiometric hydrolysis) was added. As plastisizer, 33.04 g of TEGDMA (triethylene glycol dimethacrylate) was added. In order to initialise photopolymerization, 0.01 mole of the photoinitiator Irgacure 184 was added per mole of C=C double bond. After stirring for 4 hours, ethanol was added as solvent to adjust the viscosity of the solution. All this experiments were carried out under UV and blue light exclusion in order to prevent undesired photopolymerisation. The diameter of the ZrO_2-particles, which are formed by the in-situ condensation of the ZR/MA in the MPTS matrix, was measured in the liquid phase by photon correlation spectroscopy (Laser Goniometer ALV/SP-125 #10).

Slurry for tape casting

A slurry with a solid content of 50 wt.% is made by mixing 50 wt.% ZrO_2, 6,25 wt.% TODS (12,5 wt.% respective to ZrO_2), 7,5 wt.% polyvinyl alcohol PVA 4-88 (15 wt.% respective to ZrO_2), 2,25 wt.% glycerol (30 wt.% respective to PVA) and 34 wt.% water.

RESULTS AND DISCUSSION

All experiments were based on the precipitation of ZrO_2 from zirconium propoxide as described in the experimental. The processing route is schematically shown in figure 1. As shown elsewhere by X-ray diffraction [15], the ZrO_2 nanoparticles are completely non-crystalline. The particle size of the amorphous precipitates in the process for the holographic systems was about 4 nm, as measured by photon correlation spectroscopy. The hydrothermal processing was carried out by the newly developed process shown in figure 2. In this process, the precipitated amorphous slurry was crystallized in a continuously working tubular reactor. This reactor with an inner volume of 25 liters is constructed in a modular arrangement of 27 stainless steel coils (each 7 m in length). The maximum working temperature is 500 °C. The pressure, produced by a high pressure membrane pump, can be chosen from 10 bars to 220 bars. The hydrothermal treatment time is controlled by the flow rate of the suspension. The production capacity is depending on the solid content of the suspension and the required treatment time. A water-based amorphous suspension with 10 wt. % solid content was pumped with a flow rate of 50 liters/h trough the heated coil reactor at a temperature of 250 °C. On that condition the production capacity of the reactor for zirconia is 5 kg/h. The obtained fully crystalline suspension is weakly agglomerated with a primary crystal size of 9 nm (determined by X-ray powder diffraction) and a BET-surface of 150 m^2/g. After removing of the no longer required processing additives by a washing step, the suspension was freeze dried (Christ, Epsilon 2-60) for further use. In figure 1 the flow chart of the process is shown. The core unit is the tubular reaction chamber having a total length of about 190 m (figure 2). The precipitation takes place in the mixing chamber b.

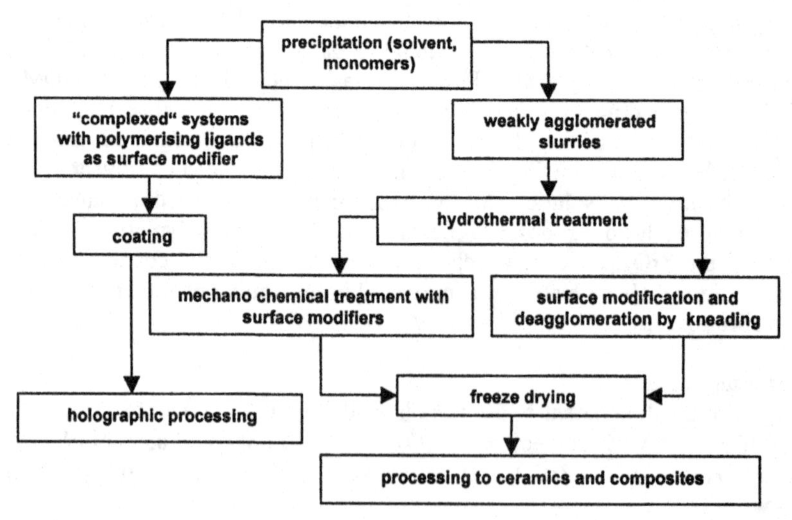

Fig. 1. Scheme of the different processing routes for ZrO_2 nanoparticles

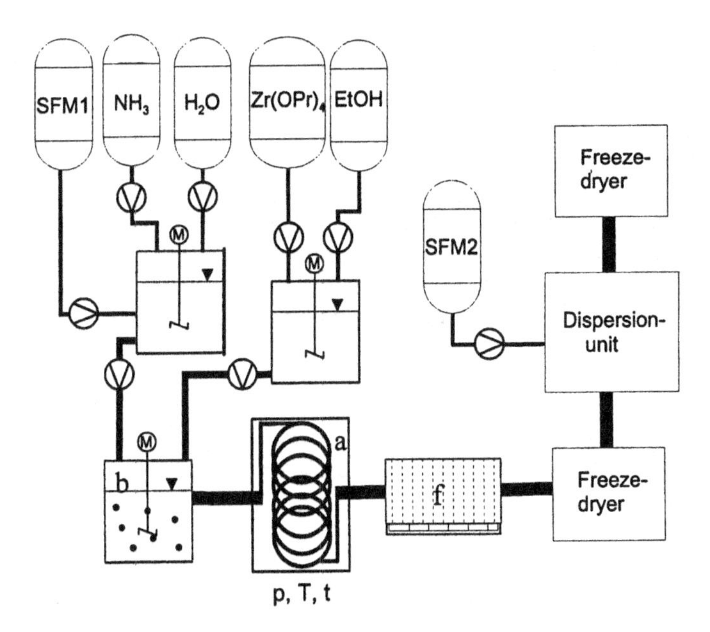

Fig. 2. Scheme of the hydrothermal process in a continuously working tubular reactor; a: reaction tubes; b: precipitation chamber; SFM 1, 2: surface modifier addition; f: filtering unit

The particle size can be manipulated by altering the processing parameters of the hydrothermal treatment. By prolongation of the reaction time the particle size was influenced as shown in figure 3 in a narrow particle size range of 7 to 10 nm. Even the process of particle growth is slow under the chosen conditions. First experiment by changing other variables in the overall reaction allowed an acceleration up to the factor of four (to be published in detail later).

Two different methods for the deagglomeration and surface modification were utilized in this study, kneading and milling. For an effective kneading process a paste able to be kneaded is needed. This means, that an optimised paste has to have a sufficiently high viscosity, high cohesion and low adhesion. If these criteria are fulfilled, as it is achieved by the device described in the experimental part, it is possible to separate the agglomerated particles due the high shear forces, which are provided by the kneader rotors. Because of the generation of heat the kneading process is limited by the cooling capacity of the kneader to avoid temperatures above the boiling point of the solvents. For deagglomeration, 1 kg of the freeze dried powder, 10-15 wt. % surface modifier and 80 ml water were filled into the kneader. After 4 hours of kneading the following particle size distribution (d_{10}, d_{50} und d_{90} values represent the volume fractions) of the zirconia powder could be achieved: d_{10} = 10 nm, d_{50} = 13nm, and d_{90}= 17 nm. The surface modified powder is fully dispersable in water and can be used e.g. for tape casting or for compounding into polymers.

As second route for deagglomeration and surface modification a ball milling process was investigated. The combination of the SMSM-concept (Small Molecule Surface Modification) [11] with the technology of wet grinding enables the production of ZrO_2 colloids from neat powders. In contrast to conventional milling additives small molecules with specific functional groups are utilized, tailoring the particle surface chemistry for appropriate processing.

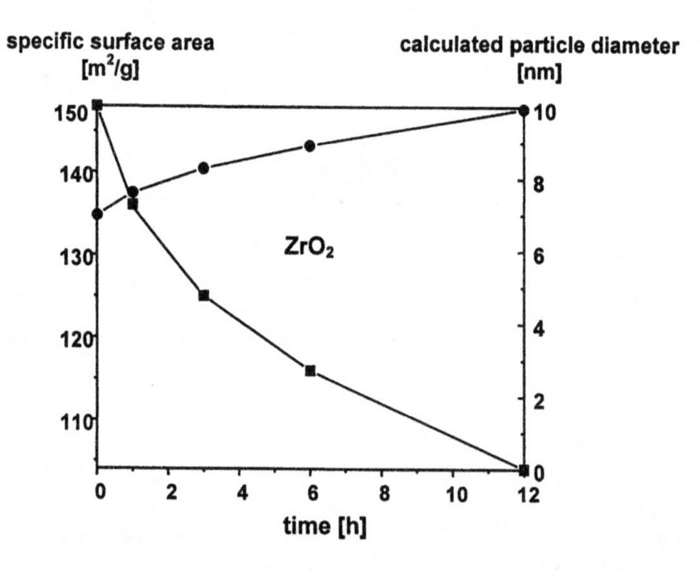

Fig. 3. Particle growth with time during hydrothermal treatment

The chemo mechanial comminution reaction in this study was carried out in a drais ball mill according to the laboratory set up shown in figure 4.

As starting material, the agglomerated zirconia, fabricated by the former mentioned hydrothermal process has been used. Water based suspensions of ZrO_2 and the surface modifiers listed in the experimental were subject to the mechanical comminution reaction leading to highly transparent and stable colloids. During this milling process high mechanical stress (shear and impact force) by the moving milling balls is introduced to the powder breaking down the agglomerates. In the present of a surface modifier strong bonds simultaneously were formed to the fresh produced surfaces. The resultant saturation of the reactive surface groups prevents the reagglomeration of the powder. For this type of reaction it is of high

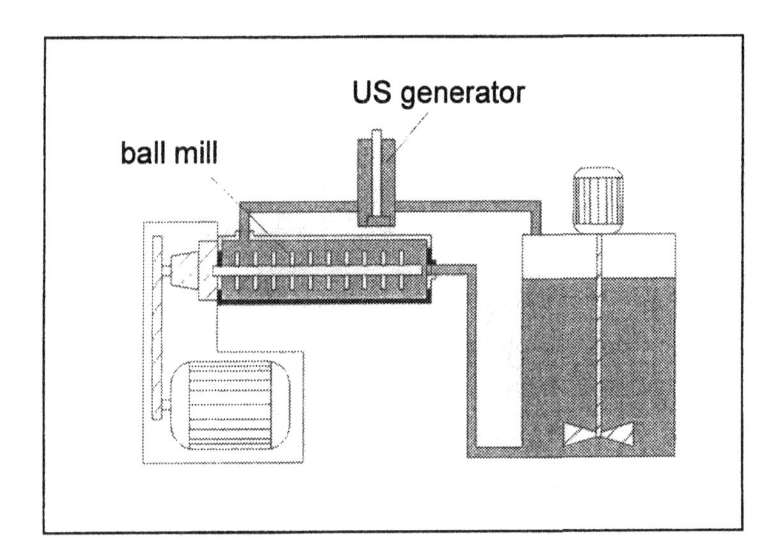

Fig. 4. Scheme of the ball mill device using an ultra sonic generator in the flow system

importance to utilize relatively small molecules, because the process is diffusion controlled. The second requirement for the surface modifier is to produce good stable bonds to the particle surface.

Surface modifiers used in these investigations are listed in table I as well as the achieved particle sizes after 4h of milling. The best result has been obtained by N-(2-hydroxyethyl)iminodiacetic acid where a d_{50} value of 11 and a d_{90} value of 14 nm has been realized. Although the values obtained from 3,6,9-trioxadecanic acid (TODS) are very reasonable. Based on the availability, most experiments have been carried out with 3,6,9-trioxadecanic acid. Still the performance of formic acid can be described as sufficient. However it was a little bit surprising, because of the very small size of formic acid we had expected a much better dispersion state. This is still under investigation. A systematic examination on the surface modifier concentration dependency of the comminution reaction was carried out for the TODS surface modifier. The concentration of TODS was varied in 5 wt.% steps from 5 to 15 wt.%. There is a clear connection between the concentration of the modifier und the results obtained by the grinding process. 15 wt.% TODS are required to accomplish the best particle size reduction during the grinding (table II).

In the case of the surface modifier TODS [15 wt.%] the chemo mechanical comminution reaction was also monitored by taking specimens of the suspension every hour. These measurements clearly indicate (figure 5) a particle size

Table I. Surface modifier and obtained particle size (d_{10}, d_{50} und d_{90}- value of the volume distribution) by the grinding process (grinding time 4h)

	Surface modifier	d_{10} [nm]	d_{50} [nm]	d [n
alkylcarboxylic acids	formic acid	12	17	4
	acetic acid	10	13	2
ethercarboxylic acids	3-oxabutanic acid	9	12	3
	3,6-dioxaheptanic acid	9	13	2
	3,6,9-trioxadecanic acid	9	11	2
amino acids	N-(2-hydroxyethyl)ethylendiamine-N,N',N'-triacetic acid	10	13	3
	N-(2-hydroxyethyl)iminodiacetic acid	9	11	1
	bicine	8	10	1
	6-aminohexanic acid	11	15	3

reduction during the grinding according to expectations reaching a minimum after 5 h (d_{10} = 9 nm, d_{50} = 11nm, and d_{90}= 20 nm).

Table II. Obtained particle size (d_{10}, d_{50} und d_{90}- value of the volume distribution) by the grinding process with 5,10, 15 wt.% TODS as surface modifier.

wt.% TODS	d_{10}	d_{50}	d_{90}
5	36	72	155
10	12	47	135
15	9	11	23

After the strong decrease of the particle size during the first hour to d_{10} = 14 nm, d_{50} = 48 nm, and d_{90}= 185 nm, practically no effective size reduction occurred after 4 h of grinding (only reduction of d_{90} value in the 5th hour from 23 to 20 nm, d_{10} and d_{50} value are unchanged). Compared to the d_{90} value, the d_{50} and also the d_{10} displayed a much weaker dependency of the milling time. The d_{10} value is to a large extent independent of the milling time and is approximately identical with the primary crystal size of 9 nm.

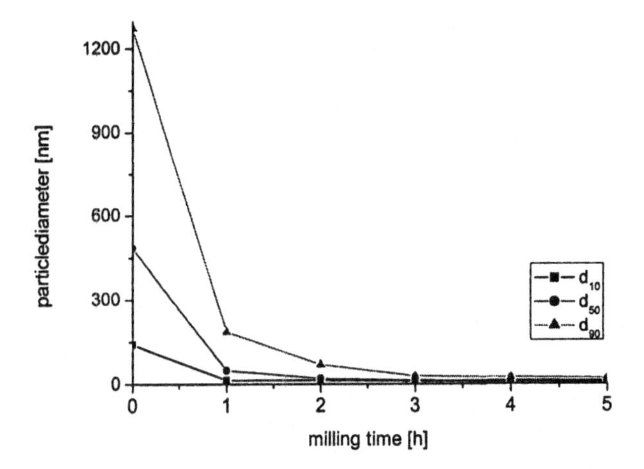

Fig. 5. Time dependence of the d_{10}, d_{50} and d_{90} values on the volume distribution

As stated before for a efficient surface modification, it is of high importance to get good chemical bonds of relatively small molecules to the surface. This is shown in figure 6 for TODS where it can be clearly seen that the CO frequency (free acid) is shifted to lower wave numbers in the bonded case. This confirms that a strong chemical bond is formed, which is necessary to maintain the surface modification in the further processing steps.

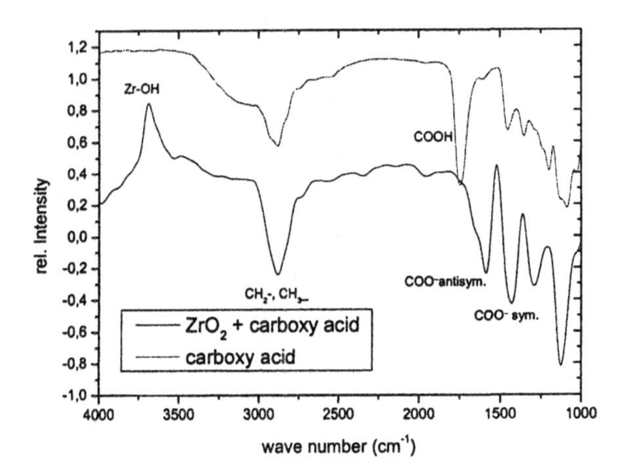

Fig. 6. DRIFT-spectroscopy of a surface modified zirconia sample (15 wt.% TODS) and the pure surface modifier (upper curve); the pure TODS is measured in transmission mode

To broaden the field of functional surface modifiers sorbic acid [10 wt.%], a unsaturated carboxylic acid was also tested. The viscosity of the water based suspension increases extremely after a short milling time so that the test could not be continued. This might be attributed to the low solubility of sorbic acid in water. To overcome this drawback the dispersion media was changed to ethanol with 1,5 wt.% water. In contrast to the former observation, no increase in viscosity took place and a comparable particle size distribution to the water based systems could be obtained after 6 hours of grinding (d_{10} = 9 nm, d_{50} = 11 nm, d_{90} = 14 nm).

Tape casting

Zirconia powders as derived from the hydrothermal processing and the subsequent surface modification have been used for the fabrication of ceramic membranes. For this reason, a tape casting process has been developed. A slurry with a solid content of 50 wt.% was made by mixing 50 wt.% ZrO_2, 6,25 wt.% TODS (12,5 wt.% respective to ZrO_2), 7,5 wt.% polyvinylalcohole PVA 4-88 (15 wt.% respective to ZrO_2), 2,25 wt.% glycerol (30 wt.% respective to PVA) and 34 wt.% water. The solid content of the resulted slurry was 50 wt.% (15 vol.%). After tape casting and subsequently drying the green density of the tape reached 79 wt.% (42 vol.%). The casted tapes have been dried and can be stored in the form of rolls. For the fabrication of the membrane, the zirconia separation layer has been laminated on top of a porous support, for example, alumina and sintered at 1000°C to a micro or nanoporous film. In figure 7, the cross section of a membrane fabricated by this technology is shown.

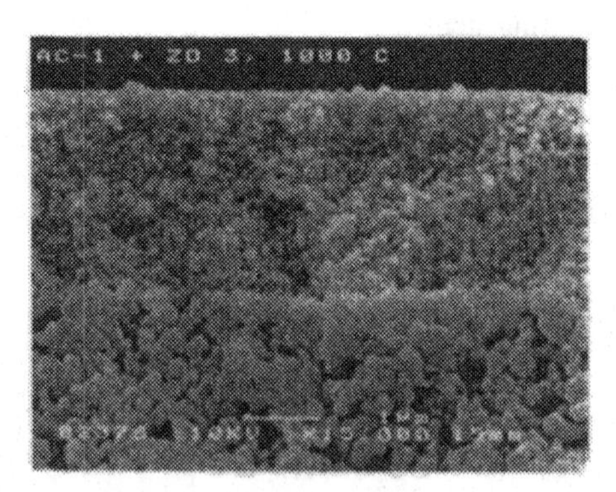

Fig. 7. SEM-micrograph of a nanoscaled zirconia membrane layer on a porous alumina support

The advantage of this type of membrane fabrication is two-fold. First, the membrane can be fabricated independently and the green membrane tape can be stored for a long time, and second, the flatness of these membranes is much higher than the conventional technology by sol-gel dip-coating. A first state about the separation behaviour is shown in table III.

Table III. Gas phase separation behaviour of the nanofiltration membrane

Temperature	H_2-Permeation	Permselectivities
25 °C	7*10-5 mol/Pa*s*m²	H_2 / Air: 3,2 (independent of the temperature)
200 °C	5,5*10-5 mol/Pa*s*m	H_2 / CO_2: 3,6 (independent of the temperature)

Polymer Compounding

The fabrication of polymer matrix compounds has been carried out by using surface modified zirconia particles by trioxadecanic acid (12.5 wt.% trioxadecanic acid on zirconia with 9 nm primary particle size). This has been used to stabilize polycarbonate in a melt compounding process. The effect on the compressive strength is shown in figure 8. The bars of 20 x 20 x 100 mm^3 in size have been put under an axial pressure of 100 MPa. As one can see, the unstrengthened polycarbonate is completely deformed whereas the polycarbonate filled with 5 wt .% of ZrO_2 does not deform at all.

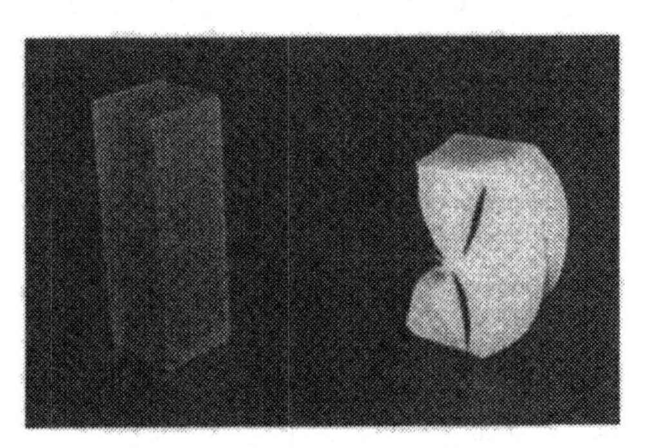

Fig. 8. Comparison of two PC bars of identical dimensions in the pressure test

The effect of the surface modifier for the homogeneity of the particle dispersion is shown in figure 9. Whereas the unmodified zirconia cannot be dispersed in the melt compounding process (using a Werner & Pfleiderer two screw extruder), the surface zirconia modified by TODS (see above) is homogeneously dispersed. The polycarbonate granules were melted in the extruder and the surface modified zirconia nanoparticles have been added the polycarbonate melt as in the conventional compounding process as a powder.

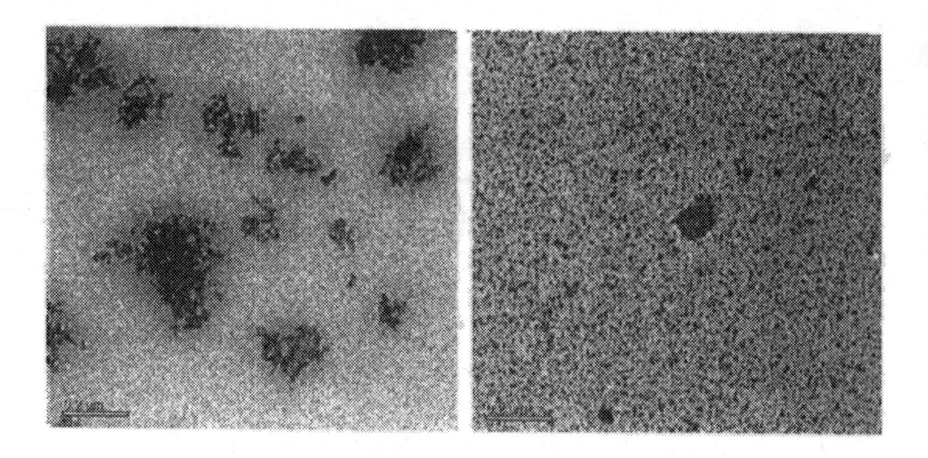

Fig. 9. Transmission electron micrograph (TEM) of unmodified (left) and modi fied (right) zirconia in the polymer

The photo induced diffusion

Amorphous zirconia nanoparticles prepared according to the experimental part, suspended in prehydrolysed MPTS, have been irradiated through a mask or by two wave mixing as described in [15]. In figure 10, the scheme of the process is shown and in figure 11, the diffractive pattern which has been obtained by the up concentration of the zirconia nanoparticles. The lines have be verified by EDX,

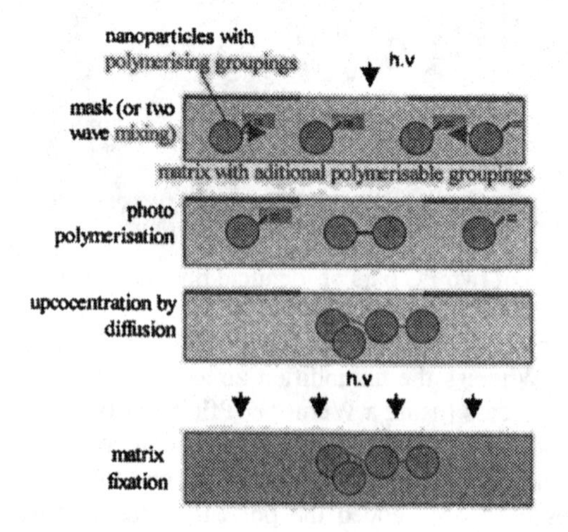

shown in figure 11. By two wave mixing process of a concentric with a planar wave, fresnel lenses have been written using this technology. In the meanwhile, an industrial process has been developed for the fabrication of light guiding films by a continuous process, where the photo induced micropaterning is carried out online after the wet web coating of the patternable nanocomposite films on a carrier plastic foil.

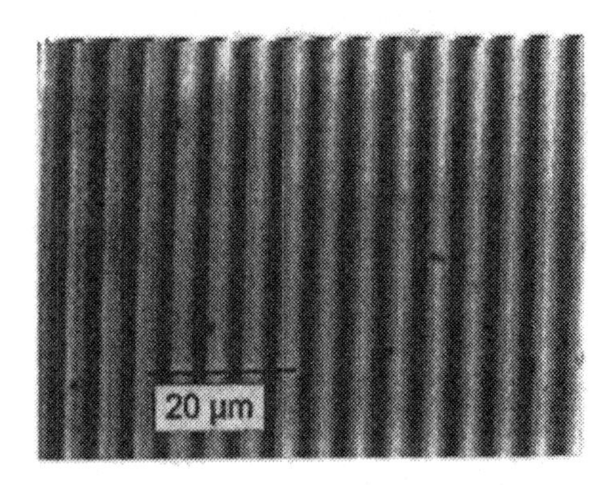

Fig. 11. Diffraction Patterns obtained by the ZrO_2 holographic process

CONCLUSION

One of the most important conclusion of this work is, that the processing of nanoparticles is strongly facilitated by the appropriate surface modification. This includes fabrication as well as processing to films and composites. Especially if optical systems are envisaged, a homogeneous dispersion is indispensable to reduce Rayleigh scattering. This process as shown with zirconia can be generalized and used also for other nanoparticles.

ACKNOWLEDGEMENT

The authors want to thank the State of Saarland, the Federal Ministry for Education, Research and Technology of the Federal Republic of Germany, as well as many industrial partners for supporting the work.

REFERENCES

[1] World Congress on Particle Technology 4 , Sydney, Australia (2002) and Literatur presented in Proceedings of World Congress on Particle Technology 4, CD-ROM(2002).

[2] NANOPARTICLES AND NANOSTRUCTURES THROUGH VAPOR PHASE SYNTHESIS, Vapor Phase Synthesis of Materials IV, UNITED ENGINEERING FOUNDATION, Tuscany, Italy (2002) .

[3] Particles 2001, Powder Conference, American Chemical Society, Orlando, Florida (2001).

[4] Mädler, L., Pratsinis,S.E. "Flame Spray Pyrolysis (FSP) for Synthesis of Nanoparticles", World Congress on Particle Technology 4, CD-ROM (2002) paper 144, Sydney, Australia, July 21-25 (2002).

[5] Stephen O'Brien, "Oxide nanoparticles: Synthesis strategy and size dependent properties", Particles 2001, Orlando, FL, U.S.A., Book of Abstracts page 195 February (2001).

[6] Andreas Gutsch,"Project House Nanomaterials - A new concept of strategic research", Particles 2001, Orlando, FL, U.S.A., Book of Abstracts page 50, February (2001).

[7] Mädler, L., Mueller, R., Pratsinis, S.E. "Synthesis of Nanostructured Particles by Flame Spray Pyrolysis", Particles 2001, Orlando, FL, U.S.A., Book of Abstracts, February (2001).

[8] Schmidt H.K. *Kona powder and particle* 1996:**14**, 92-103

[9] H. Schmidt, "Nanoparticles by chemical synthesis, processing to materials and innovative applications" Applied Organometallic Chemistry, **15** 331-343 (2001).

[10] H. K. Schmidt, "Nanoparticles for ceramic nanocomposite processing", Molecular Crystals and Liquid Crystals **353** 165-179 (2000).

[11] H. K. Schmidt, "Das Sol-Gel-Verfahren : Anorganische Synthesemethoden", Chemie in unserer Zeit **35** 176-184 (2001).

[12] C. J. Brinker, G. W. Scherer, Sol-gel science: the physics and chemistry of sol-gel processing, Academic Press, Boston (1990).

[13] D. Burgard, R. Naß, H. Schmidt, "Process for producing weakly agglomerated nanoscalar particles", U.S. Pat. No. 5,935,275, August 10, 1999.

[14] F. Tabellion, H. Schmidt, P. Müller, "Production and processing of superior nanoscaled powders" , Material-Forum, Hannover-Messe, Germany (2003).

[15] H. Krug, P. W. Oliveira, H. Künstle, H. Schmidt, "The production of fresnel lenses in sol-gel derived ormocers by holography", SPIE Volume 2288 Sol-Gel Optics III, 554-562 (1994).

THE COMMERCIALIZATION OF NANOMATERIALS

Edward G. Ludwig
Nanophase Technologies Corporation
1319 Marquette Drive
Romeoville, IL 60446

ABSTRACT

Nanophase Technologies Corporation engineers and manufactures nanocrystalline materials on a commercial scale to provide value-enhanced solutions to customers. As a nanomaterials company, Nanophase's interest is to gather core technologies that provide the capability for significant growth of revenue and profit. Our focus is to move up the value chain and supply a fully engineered material capable of meeting a customer requirement. It is this integration of nanotechnologies to provide optimally engineered solutions to a customer that produces the best value. We call this the "solution-provider" business model and the strategies that will be presented should be viewed in the light of this business model. This presentation will focus on the following: emerging trends and applications for nanomaterials, commercialization strategies for nanomaterials and integrating nanotechnologies for commercialization.

OBJECTIVE

The objective of this paper is to illustrate the commercialization of nanomaterials through approaches to the technology and to the marketplace.

We will also use examples of how Nanophase Technologies Corporation approaches the marketplace to provide clear understanding.

THE COMMERCIALIZATION OF NANOMATERIALS

Nanotechnology is the science of engineering materials at the nanometer scale. The result of the effort is often realizing novel or improved material properties, and in many cases superior performance characteristics. Nanocrystalline materials are defined as particles that are less than 100 nm in diameter. The properties of nanomaterials typically depend on composition, particle size, distribution and shape, chemical and crystal structure and surface characteristics.

The benefits of nanomaterials often include novel behavior and performance not found in larger materials. These may include particles that, due to their size, are transparent to visible light and so may be used to impart abrasion resistance, UV protection or electrical conduction in optically transparent coatings. Also, due to their size, nanoparticles can be embedded in fibers and coatings, including passing through spinnerets, to achieve benefits such as anti-microbial activity. Small size indicates high surface area, and this can. The following list represents opportunities for nanoparticles that are either commercial or potentially commercial.

- Packaging, including charge dissipative packaging for electronics
- Coatings, such as permanent UV blocking and abrasion resistant clear coatings
- Fuel and explosive additives, including nano-sized aluminum
- Abrasives, such as metal and glass polishing materials and ceria for CMP applications
- Medical and biopharmaceutical, including anti-microbial materials
- Bioanalysis and medical analysis, such as various gas-porous sensors
- Catalysis, including environmental automotive converters and chemical process catalysts
- Cosmetics, such as UV blocking formulas and special pigment systems
- Fuel cells and batteries, such as for hydrogen and oxygen transport in solid oxide fuel cells

To succeed commercially, we must create value enhanced solutions for our customer's problems. Nanoparticles produced and surface treated on a commercial scale have demonstrated unique performance characteristics in several important fully commercial applications.

1. Nano zinc oxide is used to attenuate UV in optically clear sun screen products and is used to absorb odor in various personal care

products such as foot powders.

2. Nano ceria is used in catalytic converters, chemical mechanical planarization of semi-conductor wafers and for polishing photomasks, rigid memory disks and specialized optics.

3. Nano alumina is used in clear wear layers for vinyl and wood laminate flooring to impart improved scratch resistance.

New applications areas are also being investigated and the following three examples are representative of such activities.

1. Nano zinc oxide is being tested for effectiveness as an anti-microbial ingredient in fibers, hard surface cleaners, filters and textiles.

2. Nano-sized mixed rare earth metal oxides are being evaluated for environmental catalysts for both gasoline and diesel applications and for use in fuel cell applications.

3. Nano-sized doped tin oxides are being evaluated for use in charge dissipative packaging for electronics and other applications.

We believe there are four strategic goals that are critical to the success of the business. Each of these goals must have the full commitment of the senior management team and are the focus of ongoing effort on a daily basis. We focus on maintaining a technical leadership position in the nanocrystalline materials area while delivering innovative solutions to customers' problems, building a sustainable infrastructure and driving revenue growth.

Customer satisfaction is directly tied to innovation. Securing technological innovation requires an intellectual property strategy that provides protection using a combination of company-owned patents, patents developed in conjunction with partners, licensing of technology and specific trade secrets.

Using Nanophase as an example, our PVS® process requires that the anode be an electrically conductive anode that is consumed in the process. More recently, we have invented a patent-pending plasma technology called NanoArc Synthesis™ that feeds powders and can effectively open the entire periodic table to nanomaterial production. We also have strong protection around our surface modification and dispersion technology. Our IP strategy includes filing for specific applications patents, in conjunction with our partners, as appropriate.

Our business model focuses on providing a solution, not on just selling a product. We focus on co-development projects where there is an unsatisfied need, an internal champion and a sense of urgency to get a product to the market. Rather than just selling nanoparticles, we strive to add value through, for example, modifying the surface characteristics or offering the products in a pre-dispersed form, where the product is broken down into primary particles. Thus, we see ourselves as providing a solution rather than supplying a product or material.

There are three key and important results of this approach:

1. We achieve higher value because we have moved up the value chain.

2. The time to market is decreased because we have supplied the product in a readily useable form.

3. Development risks are reduced because our scientists are working directly with our partners' scientists.

To be certain that we are in step with our partners, we take EH&S (environmental, health and safety) considerations very seriously. Some suggestions on ways to demonstrate a serious EH&S program are provided here. The workplace must be safe and considerable investments must be made in employee training. For example, Nanophase has gone in excess of 450,000 hours of operation without a lost time injury. All waste permits should be current and the company should be in complete compliance. Don't pass the buck – third parties responsible for waste disposal should be audited frequently. Having a good reputation with OSHA (Occupational Safety and Health Administration) and the EPA (Environmental Protection Agency) is important, and ongoing activities for waste reduction and recycling are critical. Most companies, and especially large companies, will consider EH&S carefully when evaluating and selecting a partner and supplier.

Before a customer will trade their money for a product, be it dollars, euros or yen, performance must be clearly demonstrated. We must either enable a process or improve a product that on a systems cost basis gives our customer a competitive advantage in the marketplace. In the nanotechnology field, we must sell performance!

Nanomaterials are almost always more expensive than the corresponding formulation in bulk chemical form on a per pound basis. Therefore, there must be a benefit in terms of usage level, properties, performance or some combination of these that provides appropriate value to the customer and end-user. If such benefits cannot be determined or quantified, the application may not lend itself readily to nanotechnology.

We've all heard the hype about nanotechnology, from quantum dot computers to nanorobots that are self-replicating to a host of medical applications. The more complex the application, typically the longer it will take to reach commercial fruition.

We are building a financially strong business by focusing predominately on the moderate applications that we expect will realize revenue in the 12 to 18-month timeframe. This provides two important benefits to us:

1. The technology is validated by providing competitive nanomaterials into applications such as vinyl flooring, catalytic converters, transparent functional coatings and personal care products.

2. We are able to build critical skills in engineering, surface treatment, commercial scale-up and infrastructure that will allow us to tackle more difficult applications in the future.

Getting a product into the marketplace is critical, and it is important to choose the best route. Moderate applications probably would involve an established channel to market, especially if the nanomaterials can solve a market need. We also search for applications in which nanomaterials offer benefits not readily achievable by micron-sized products and applications where value is added, typically resulting in a market pull.

There are a number of factors that must be considered when determining the entry point on the value chain. Moving higher up the value chain can result in perceived higher financial returns, but it may come at the price of higher entry barriers, longer time to market and slower or lower market penetration.

Our experience has been that we have had the most success by partnering with established industry leaders who supply the bulk chemical versions of the nanomaterials into specified applications. These companies are often very large, sometimes hundreds of times bigger than we are, control the distribution channel and have the contacts and inroads into formulators and end-users.

Nanotechnology has already been successful in a variety of end-use applications, many of which have already been mentioned. In order to insure that we are making the best use of our resources, we ask a series of questions:

1. Is this a market pull or a technology push? We like market pulls.

2. Is there a clear unsatisfied need and an internal champion?

   We then attempt to better understand the opportunity:

1. What is the revenue potential?

2. What is the technical complexity of the project – is it feasible and do we have a chance for success?

3. What are the marketing opportunities, for example, the scope of the horizontal market applications outside the field of use?

4. Can we develop a partnership relationship with the prospective customer to minimize risk, enhance the channel to market and improve the speed to market?

As an example of what I mean by a vertical and horizontal marketing approach, let me use our zinc oxide business. We have developed a long term mutually exclusive agreement with BASF to supply ZnO for human sunscreen applications. Developing the capabilities to surface treat and manufacture ZnO under FDA mandated cGMP while delivering hundreds of metric tons per year opens the door to expand sales of ZnO to markets outside the sunscreen field of use. We look for horizontal market applications with zinc oxide that will generate revenue opportunities that fit our 12 to 18 month time to market. Currently, there are four promising areas in which we are exploiting our zinc oxide manufacturing expertise:

1. Personal care products other than sunscreens (where odor absorbing and anti-inflammatory characteristics have utility), also through BASF.

2. Electronics applications where our product is undergoing testing.

3. A component in clear coatings for UV attenuation in several different industrial applications.

4. Anti-microbial applications for hard surface cleaners, filters, fibers, textiles and paper applications.

Now we need to explore commercial operations. To grow, focus must be placed on support for new product development with a clear route to scale up new products to commercial production. To reach profitability, improvement efforts in process economics and asset utilization must be ongoing. To help ensure customer satisfaction and product consistency, process control measures must be put in place. To be successful, we must be able to scale up the new products under economies of scale that let us achieve profitability.

To partner with large companies, in addition to the EH&S considerations mentioned previously, we have to be at the top of our game in quality. ISO certification, and especially compliance with the new ISO 9001:2000, is very important. Larger companies will pay more attention to a supplier that is ISO compliant. In order to participate in the pharmaceuticals arena, which includes active ingredients in personal care applications, registering as a cGMP manufacturer for bulk production of pharmaceutical-grade products is a requirement. In such cases, the FDA and various customers will want to perform frequent audits.

Having a state of the art QA lab is necessary to back up quality claims and handle customer inquiries and complaints. A focus on continuous improvement through the use of process improvement teams is a good idea and operating under a customer driven quality approach is advisable. As an example, we are able to make thousands of tons of products per year and have achieved 99+% on-time delivery with our customers. Operating two separate manufacturing plants with a pilot plant and scale-up facilities, there is complete capability for particle coating, dispersion, blending, packaging, warehousing and distribution. If you claim to have it, your customers and partners will want to see it.

As an example of partnering to jointly deliver solutions to the marketplace, BASF required a hydrophobic coating on zinc oxide that is compatible with electrolytes. Our first coated product allowed for less than a 2% concentration in a sunscreen formulation, but we developed a patented coating composition that allows our zinc oxide to be used in formulations at concentrations greater than 16% without difficulties. This is a very important commercial product for us today. But it didn't come easy!

While success has been realized with our zinc oxide product line, it did not happen without a lot of hard work and attention to details. Over the past four years, production rate and process capability improvements have driven a considerable reduction in manufacturing cost and an increase in manufacturing capacity. The learning gained from the efforts made in the zinc oxide product line is directly transferable to a variety of other products and technologies.

In the CMP and glass polishing areas, we have partnered with Rodel and Schott. As an example of specific surface modification and dispersion technology, we make a dispersion for Rodel that is at a pH of 5 and a particle zeta potential of +45 mV, while a similar product is prepared for Schott, but the dispersion has a pH of 8 and the particle zeta potential is –45 mV. Both products are prepared while maintaining the tight particle size distribution. It should be noted that the particle size measured in aqueous media by light scattering techniques is the hydrodynamic radius, and so the ceria, with a dry average particle size of 30 nm is measured as roughly 100 nm in aqueous media.

CONCLUSION

The keys to success listed here are applicable to just about any chemical business, but are especially critical for nanotechnology because we are selling performance.

- There must be a clear commitment to EHS and Q – to have a good reputation in these areas, things must be done right and done right the first time.

- The key word in nanotechnology is technology – to be successful, there is a need for top-notch technical people and a management team that understands how to market technical products.

- Success means protecting what you know and using it to build a business – focus on strengths and strive to be on the cutting edge.

- Move up the value chain to deliver a solution that the customer and market need, don't just sell a product.

- Strive to get long-term relationships with leaders who control the channel to market – if done properly, this should maximize penetration and minimize time to market.

- Focus on emerging markets and applications, but be sure that focus is reasonable in terms of technical feasibility and time to market, and that your company's infrastructure can support the effort.

PANEL DISCUSSION: COMMERCIALIZATION OF NANOMATERIALS

S. W. Lu

PPG Industries, Inc., Glass Technology Center, P. O. Box 11472, Pittsburgh, PA 15238-0472

Summary of panel discussion held in conjunction with Symposium 7, "Nanostructured Materials and Nanotechnology" during the 105th Annual Meeting of the American Ceramic Society, Wednesday, April 30, 2003, 12:00 - 1:00 pm in Nashville, Tennessee

INTRODUCTION

In a relatively short time, the field of nanostructured materials has expanded from a novel area of research to a technology with a significant and rapidly growing commercial sector. For the first time in a series of ACerS symposia related to nanotechnology, a panel discussion "Commercialization of Nanomaterials" was held in conjunction with Symposium 7, "Nanostructured Materials and Nanotechnology" during the 105th Annual Meeting of the American Ceramic Society, Wednesday, April 30, 2003, in Nashville, Tennessee. Five scientists and business leaders who are involved with the commercial development and applications of nanostructured materials were invited as panelists to discuss important issues related to the commercialization of nanomaterials. The panel discussion was highlighted due to the fact that all panelists would present their invited presentations of their latest developments in a session "Industrial Development and Application of Nanomaterials" right after the panel discussion. Approximately one hundred people attended and listened to the hour-long discussion.

The five invited panelists were H. K. Schmidt, Managing Director of Institut für Neue Materialien (INM), GmbH, Germany; K. -L. Choy, Founder and Technical Director of Innovative Materials Processing Technologies (IMPT) Limited, and Professor, University of Nottingham, England; E. G. Ludwig, Vice President for Business Development of Nanophase Technologies Corporation; G. Skandan, Chief Operating Officer of NEI Corporation (formerly Nanopowder Enterprises Inc.); and R. M. Laine, Chief Technology Officer of TAL Materials,

Inc. and Professor, University of Michigan. The panel discussion was organized and moderated by S. W. Lu of PPG Industries, Inc. The topics were published in advance in the meeting program. Mark De Guire, lead organizer for Symposium 7, and Noel Vanier, PPG Industries, Inc., took notes on the discussion. All panelists were given the chance to review the present manuscript for accuracy prior to publication.

NANOMATERIALS AND THEIR APPLICATIONS

The panel discussion started with the pre-selected topic, "What are the most commercially important nanomaterials and their applications? What areas are predicted to become important soon?" The panel listed several important areas of nanomaterial applications based on their own experiences. Using technology protected by over 30 patents and applications together with proprietary technology, Nanophase Technology Corporation, a leading nanotechnology company based in Romeoville, Illinois, produces nanocrystalline materials that are currently used in sunscreens, personal care formulations, abrasion resistant transparent coatings for floor coverings, and catalytic converters. In addition, they are developing several new emerging applications for nanocrystalline materials in multiple markets. "We make something that can't be done with micro-sized particles," said Ludwig. Current products of Nanophase include NanoTek® series oxide nanocrystals such as aluminum oxide, antimony doped tin oxide, cerium oxide, copper oxide, indium tin oxide, iron oxide, titanium dioxide, yttrium oxide, and zinc oxide. Several important applications are specialty ceramics using nanocrystalline aluminum oxide and zirconia-toughened alumina, transparent coatings for attenuating IR and UV radiation, static dissipation and conductive films fabricated from antimony or indium doped tin oxides, catalysts for chemical and environmental applications, such as cerium oxide and iron oxide, and materials to improve topical healthcare products such as nanocrystalline zinc oxide.

Schmidt emphasized that chemical nanotechnology through surface modification was very important for industrial development and applications of nanomaterials. Under his leadership, INM has developed various nanomaterials, nanocomposites, and their applications via chemical nanotechnology in which nanoparticles can be chemically tailored. INM has shown high potential for cost-effective manufacturing of nanomaterials through chemical nanotechnology. By tailoring the surface chemistry of the nanomaterials, agglomeration can be inhibited and novel functions can be obtained. Developed nanomaterials include magnetic oxide (Fe_2O_3), functional oxide particles (e.g. TiO_2, SiO_2, talcum, CeO_2, $BaTiO_3$, lead-zirconate-titanate), and inorganic-organic nanocomposites. Various applications of nanomaterials have been developed, e.g., transparent electrically conductive coatings, hard coatings, photochromic films, photocatalytic films, antireflective coatings, dirt-repellant coatings, gene and drug targeting and for therapeutical purposes (e.g. tumor therapy). Schmidt also stated that for the commercialization in upcoming areas, the material development alone is insufficient. The application processes have to be developed further to meet different needs of customers.

Skandan told the audience that functional nanomaterials offered more value than non-functional nanomaterials, where nanomaterials are an enabling form of an application. Such functional applications would be extraordinary, where the nanomaterial, through its structure and composition, offers a novel functionality, e.g. Li^+-ion interaction. In addition, nanocomposites offer higher value than nanoparticles themselves. Therefore, NEI has focused on functional nanomaterials such as lithium titanate, lithium manganese oxide, and lithium vanadium oxide nanopowders for Li-based rechargeable battery electrodes, high purity alumina, magnesium oxide, yttrium oxide, and cerium oxide nanopowders and transparent nanocomposite coatings to improve scratch resistance.

On the other hand, Choy pointed out that nanomaterials have important applications in cosmetic and personal care markets. Choy has developed an electrostatic spray assisted vapor deposition (ESAVD) process to fabricate nanocrystalline powders and nanostructured coatings. Current nanostructured materials and their targeted applications using ESAVD are indium doped tin oxide coatings for flat panel displays, tin oxide coatings for sensor applications, porous Pd/Al_2O_3 films for catalysis applications, titanium dioxide and zinc oxide nanoparticles for cosmetics applications, and yttrium stabilized zirconia (YSZ) for thermal barrier coatings.

Laine felt that the most important nanoparticles were carbon black and fumed silica, which have been in the market for many years. Other important nanomaterials include titanium dioxide for photocatalytic application, aluminum oxide for paints, cerium oxide and silicon dioxide for chemical mechanical polishing, and value-added nanomaterials in bioassay devices and drug delivery. By employing a patented flame spray pyrolysis process, TAL Materials, Inc. produces mixed metal oxide nanomaterials such as TiO_2/Al_2O_3 solid solutions and nanocomposites for photocatalysis, doped yttrium-aluminum-garnet for photonic applications, YSZ for dental, sensor, fuel cell and structural ceramic applications, lanthanum manganate and gadolinium doped ceria for solid oxide fuel cell, and cerium/zirconium solid solution oxides for catalytic technologies.

CUSTOMERS FOR NANOMATERIALS

With so many important nanomaterials and their applications listed previously, there must be a lot of buyers or users of nanomaterials. The next topic for discussion was, "Who are the current and prospective customers for nanomaterials?" Laine predicted that there would be a need from electronic device companies in the near future since the shrinkage of feature sizes in electronic devices will require nanomaterials. A good example is multilayer ceramic capacitors (MLCCs). Current MLCCs are made from micro-sized metallic particles and barium titanate oxide ceramic powders. These raw materials wouldn't work for thin MLCCs whose total thickness would be one micrometer or less. In this case, only nano-sized metal particles and barium titanate oxide nanocrystals will be considered as raw materials for thin MLCCs. Ludwig, on the other hand, said that usually the first customers would be those who already supply macro-sized powder products in the market. Nanophase provides nanomaterial products to some large companies such as BASF and Rohm & Haas. The company is also continuously looking for new customers and new markets, such as nanomaterials for catalysis converters. Schmidt emphasized customer needs, "Customers don't care about nanoparticle size. They care performance and an improved product. Customers often can't process nanomaterials into their products. Nanomaterial suppliers will need to do this for them." He predicted that large companies would eventually gain this capability in the next five years. Skandan said that since large companies have their own internal nanotechnology programs, small nanomaterial companies could provide complementary products that would feed into what large companies were doing.

CURRENT LIMITATIONS OF NANOTECHNOLOGY

Although nanotechnology companies had great technological successes from starting-up, there are still tremendous hurdles that need to be overcome for future growth. Specifically, "What currently limits the commercial growth of nanomaterials?" and "How can those limitations best be addressed?" To these questions, the panelists had different points of view. Choy believed that health and safety issues related to nanomaterials were current problems for future growth. For example, companies must safely handle these nanomaterials, which are easily airborne, in order to ensure employee health and environmental safety. On the other hand, Skandan suggested that lack of understanding of cost structure of nanomaterials manufacturing was an impediment to the current profits and future growth of nanotechnology companies. This may be a common phenomenon for all nanotechnology start-ups. The cost structure changes continuously while the technology is being developed, scaled-up, and expanded until the point of full-capacity manufacturing. As a result, most costs are due to research and development, capital investment, and trials-and-failure, etc. Ludwig agreed that the high cost of nanomaterials could hurt nanotechnology companies. To this end, either the nanomaterials may be priced too high for a market, or the profit may be too little to make it worthwhile. The key to this problem, as practiced by Nanophase, is to provide value-enhanced solutions to customers on a commercial scale, e.g. to supply more readily marketable dispersions. The Nanophase Romeoville facility is focused on surface treatment and dispersions of nanomaterials for customers. Both Laine and Schmidt discussed the need to educate the public about nanostructured materials and nanotechnology. Even though the Lycurgus cup made from glass impregnated with gold nanoparticles can be dated from the 4th century A.D., nanotechnology is relatively new. The general public lacks knowledge about nanostructured materials, nanotechnology, and their applications. A key solution to address this problem is to educate the public through media and publications, and to teach customers and their engineers about the advantages of using nanoparticles in general as well as how to use specific nanomaterials in particular.

SCALE-UP OF NANOMATERIALS PRODUCTION

In an overwhelming reply to the question, "Can nanotechnology be scaled up effectively from lab to production?" all of the panelists agreed that there were no major technical problems in scaling up nanotechnology from lab to production. "If nanotechnology can't be scaled up to production, we won't be around in a few years from now," said Skandan, whose company is commercializing a variety of nanomaterials technologies. "Technical problems are a minor issue for nanotechnology firms. Collaboration between nanomaterials firms and customers is necessary in order to meet customer's specific needs," he added. Ludwig shared the same view, "Nanophase is a publicly traded company established in 1994 to commercialize a nanotechnology called physical vapor deposition (PVD). The PVD process has been scaled up very well since its start. However, the real challenge is selling nanomaterial products."

"Scale-up is not the bottleneck," added Schmidt. "It will happen if the market materializes. The most important thing in the commercialization of nanomaterials is a market for value-added nanomaterials." Laine briefly explained their nanomaterials process, "TAL Materials, Inc. has exclusively licensed a patented flame spray pyrolysis (FSP) process developed at the University of Michigan. TAL uses inexpensive raw materials for nanomaterial manufacturing at a large-scale production rate. The FSP process is inherently capable of kiloton per year. Scale-up has gone well so far."

MARKET FOR NANOMATERIALS

Market! Market!! Market!!! All panelists believed that market of nanomaterials was the most critical success factor for nanotechnology companies. "How large is the market for nanomaterials today?" and "In five years?" The answer varied from panelist to panelist. Laine discussed that the existing markets were already huge for value-added products based on nanomaterials, especially for biomedical assay applications. Ludwig told the audience that his company was focusing more on value-added nanomaterials markets, which were 0.6 million dollars in 2002 and would be 0.9 million dollars in 2005. The market is predicted to be as much as a trillion dollars from 2013 to 2018. Sixty to seventy percent of the markets by volume are oxide nanomaterials. These include applications in life sciences and electronics. Schmidt separated the markets for nanomaterials into two parts: markets for nanomaterials themselves, and markets for products whose value was added by nanomaterials. He believed that the markets for nanomaterials alone could reach a billion dollars. However, the market for products whose value is increased by nanomaterials is predicted to be much greater. The size of this market is very hard to estimate. Skandan was skeptical about current market forecasts, "It is too early to estimate overall market numbers. The market will grow from niches."

HEALTH AND SAFETY ISSUES

With the widespread research and development of nanotechnology, especially industrial manufacturing of nanostructured materials, the toxicity of nanomaterials becomes an issue for public health. "What specific handling and quality control issues related to nanomaterials do you face?" was the question from the audience. Ludwig replied, "Using state-of-the-art manufacturing facilities registered to ISO 9001 and compliant with the cGMP requirements of the U.S. Food and Drug Administration, Nanophase has done very well to handle nanomaterials in order to stay in business." Laine affirmed that employee safety is a top priority. However, we all inhale nano-sized particles in the form of airborne dust daily without adverse effects. Schmidt emphasized that hazardous airborne conditions can be eliminated in principle with liquid process in synthesizing nanomaterials, and toxicological problems can be minimized by precursor selection.

ACKNOWLEDGMENTS

The symposium organizers thank the panelists for their participation, and gratefully acknowledge the NSF Particulate Materials Center (Prof. James Adair, director) at the Pennsylvania State University for financial support of this symposium. S. Lu would like to thank Cheri Boykin, PPG Industries, Inc., for proofreading.

KEYWORD AND AUTHOR INDEX

9 781574 982039